Lecture Notes in Mathematics

A collection of informal reports and seminars
Edited by A. Dold, Heidelberg and B. Eckmann, Zürich

296

Probability and Information Theory II

Edited by
M. Behara, McMaster University, Hamilton, Ontario/Canada
K. Krickeberg, Universität Bielefeld, Bielefeld/BRD
and Université René Descartes, Paris/France
J. Wolfowitz, University of Illinois, Urbana, IL/USA

Springer-Verlag
Berlin · Heidelberg · New York 1973

AMS Subject Classifications (1970): 60G35, 60G99, 62M15, 62B10, 62C05, 62C25, 94A15

ISBN 3-540-06211-4 Springer-Verlag Berlin · Heidelberg · New York
ISBN 0-387-06211-4 Springer-Verlag New York · Heidelberg · Berlin

Offsetdruck: Julius Beltz, Hemsbach/Bergstr.

PREFACE

Interest in Probability and Information Theory, Proceedings of the International Symposium at McMaster University, Canada, April, 1968 (Lecture Notes in Mathematics, vol. 89 (1969)) was rather widespread. The deteriorating financial situation, on the other hand, made it impossible for us to organize another symposium in Probability and Information Theory. In order to keep interest in the field alive and in view of the success achieved by the first volume, the second volume has been prepared.

This volume contains papers based on invited lectures given by the authors at seminars sponsored jointly by McMaster University and the University of Montreal.

It is a great pleasure to acknowledge the cooperation and support of Professor N.C. Giri of the University of Montreal for the organization of the seminars. Our thanks go to Professor T. Husain, Chairman of the Mathematics Department at McMaster and Professors R.G. Lintz and B.J.W. Müller for their overall support. We gratefully acknowledge major financial support from the National Research Council of Canada. We are greatly indebted to Professor A.N. Bourns, F.R.S.C., President of McMaster University for his keen interest in these projects.

Our special thanks go to Professor M. Kac of Rockefeller University for his valuable advice on editorial matters.

The Editors

TABLE OF CONTENTS

UNIFORM FLOWS IN CASCADE GRAPHS

Alfred Rényi

§ 1. Cascade graphs

We shall call a directed graph G, having a finite or denumerably infinite number of points, a cascade graph, if it has the following properties:

a) There is in G a point $\underline{a}_0$--called the source--such that for any other point $\underline{a}$ of G there is in G a directed path from $\underline{a}_0$ to $\underline{a}$.

b) For any point $\underline{a}$ of G each directed path from $\underline{a}_0$ to $\underline{a}$ has the same length $r(\underline{a})$--called the rank of $\underline{a}$.

c) The number of points of G having rank k is finite for every $k \geq 1$.

Wo shall denote the set of all points of G by V and the set of all points of G having rank k by V_k $(k=0,1,2,\ldots)$. The source has clearly the rank 0 and it is the only point with this property: Thus V_0 is a one-element set containing the element $\underline{a}_0$ only, i.e., $V_0 = \{\underline{a}_0\}$. As the supposition b) has to hold for $\underline{a} = \underline{a}_0$ also, it follows that there is no directed cycle in G containing $\underline{a}_0$. It follows further that if there is in G an edge from the point $\underline{a}$ to the point $\underline{b}$ then $r(\underline{b}) - r(\underline{a}) = 1$, i.e. every edge starting from a point in V_k leads to a point in V_{k+1} $(k=0,1,\ldots)$. Let us denote by $d(\underline{a})$ the number of edges of G ending at the point $\underline{a}$, i.e. the indegree of $\underline{a}$, and by $D(\underline{a})$ the number of edges of G starting from $\underline{a}$, i.e. the outdegree of $\underline{a}$. Clearly both $d(\underline{a})$ and $D(\underline{a})$ are finite for every $\underline{a} \in V$. We call a point $\underline{a} \in V$ an endpointof G if $D(\underline{a}) = 0$. We shall denote the set of endpoints of G by E and the set of endpoints of rank k of G by E_k. For any finite set A let $|A|$ denote the number of elements of A. We put $N_k = |V_k|$ and $M_k = |V_k - E_k|$, i.e. N_k

denotes the total number of points of rank k, and M_k the number of those points of rank k, which are not endpoints. We shall denote further by R the maximum of $r(\underline{a})$ for $\underline{a} \in V$ if V is finite and put $R = +\infty$ if V is infinite. Clearly if $M_k = 0$ for some value of k then $N_m = 0$ for $m > k$ and R is finite, and conversely.

Let us consider now some examples of cascade graphs.

Example 1. If G is a rooted directed tree in which all edges are directed away from the root, and $D(\underline{a})$ is finite for every point $\underline{a}$ of G, then G is a cascade graph, its source being the root of the tree. Conversely if in a cascade graph G, $d(\underline{a}) = 1$ for all points $\underline{a}$ different from $\underline{a}_0$ (for which of course $d(\underline{a}_0) = 0$) then G is a rooted directed tree in which all edges are directed away from the root.

Example 2. Let S be a finite set. Let the points of the graph G be all subsets of S and connect $a \subseteq S$ with $b \subseteq S$ by an edge (directed from a to b) if and only if b is obtained from a by omitting one of its elements. The graph G thus obtained is a cascade graph and for every $a \subseteq S$ one has $r(a) = |S| - |a|$.

Example 3. Let S be a finite set and let the points of the graph G be all non-negative integral valued functions defined on S. If f and g are two such functions, draw an edge from f to g if and only if $g(x) \geq f(x)$ for all $x \in S$ and $\sum_{x \in S} (g(x) - f(x)) = 1$. In this way we get a cascade graph and the rank of a function f is $r(f) = \sum_{x \in S} f(x)$.

Let us call a cascade graph simple, if it does not contain any infinite directed path. In a simple cascade graph to every point $\underline{a}$ there corresponds a nonempty set $T(\underline{a})$ consisting of those endpoints of the graph which can be reached from $\underline{a}$ by a directed path. We call $T(\underline{a})$ the target-set of the point $\underline{a}$. If there is a directed path from $\underline{a} \in V$ to $\underline{b} \in V$ then $T(\underline{b}) \subseteq T(\underline{a})$.

A subset A of the vertices of a cascade graph G is called an antichain, if for any two points $\underline{a} \in A$ and $\underline{b} \in A$ there does not exist in G a directed path from $\underline{a}$ to $\underline{b}$. An antichain is called saturated, if it is not a proper subset of another antichain. An antichain is called a blocking antichain, if any directed path from the source to an endpoint, and every infinite directed path starting at the source, passes through a (unique) point of the antichain. Clearly every blocking antichain is saturated but a saturated antichain is not necessarily blocking. For instance in Example 2, let S be the set $S = \{1, 2 \ldots n\}$ where $n \geq 3$, and let the antichain A consist of the two sets $\{1\}$ and $\{2, 3, \ldots n\}$. Then A is saturated, as every subset of S not containing the set $\{1\}$ is a subset of the set $\{2, 3, \ldots, n\}$ but it is not blocking. If the cascade graph G is a rooted tree (see Example 1) then a saturated antichain is always blocking if G is finite, but not necessarily if G is infinite. (See the following Example.)

Example 4. Let the points of the graph G be all finite sequences, each term of which is one of the numbers $0, 1, \ldots, q-1$ where $q \geq 2$; the empty sequence is also a point of G. Let there be in G an edge from the point $\underline{a}$ to the point $\underline{b}$ if the sequence $\underline{b}$ is obtained from the sequence $\underline{a}$ by adding to the end of the sequence $\underline{a}$ one more digit, i.e. one of the numbers $0, 1, \ldots, q-1$. In this way we obtain a cascade graph, which is a tree, and has no endpoints. Let us take $q = 2$ and let the antichain A consist of the sequences 0, 10, 110, 1110, ... A is clearly a saturated antichain, but it is not blocking as it does not block the infinite path from the source (this being the empty sequence) leading through the points 1, 11, 111, ...

In a simple cascade graph the set of all endpoints is a blocking antichain. In any cascade graph the set V_k is an antichain and if there are no endpoints of rank $< k$, then V_k is a blocking antichain.

Let us write, for any two pointa $\underline{a}$ and $\underline{b}$ of a cascade graph, a < b if there is a directed path from a to b. G is a partially ordered set* with respect to the order relation < , but not necessarily a lattice. The following example shows that cascade graphs which are lattices have remarkable properties.

Example 5. Let $\mathcal{G}$ be a finite combinatorial geometry (see H. Crapo and G. C. Rota [1]). Let G be the graph, the points of which are the flats of $\mathcal{G}$ and there is an edge in G from the flat a to the flat b if and only if b covers a. Then G is a cascade graph, which is a lattice, and the rank of any flat a in G is the same as in $\mathcal{G}$. Let us remove from the cascade graph G its unique element with maximal rank: We obtain again a cascade graph, the endpoints of which are the copoints of $\mathcal{G}$. This cascade graph has among others the following remarkable property: The target sets corresponding to different points are different, and the target sets of points having the same rank form a Sperner-system (i.e. none of them contains any other as a subset.)

If a is any point of the cascade graph G we denote by Γa the set of all endpoints of edges starting at a. If A is any set of points of A we denote by ΓA the set of those points which are the endpoints of at least one edge starting at a point in A, i.e. we put $\Gamma A = \bigcup_{a \in A} \Gamma a$. If b is any point of G we denote by $\Gamma^{-1} b$ the set of those points a for which $b \in \Gamma a$.

§ 2. Random walks on a cascade graph

Let us assign to each edge ab (from a to b) of a cascade graph G a non-negative number w(a, b) such that

*What we call a cascade graph is, considered as a partially ordered set, a graded partially ordered sets: See Birkhoff [2] and Klarner [3], where the graded partially ordered sets with a given maximal rank and given number of points are counted.

$$\sum_{\underline{b} \in \Gamma \underline{a}} w(\underline{a}, \underline{b}) = 1 , \tag{2.1.}$$

for all vertices $\underline{a}$ of G which are not endpoints. Such a function w(a, b) defines a (Markovian) random walk on the edges of G as follows: The random walk starts always from the source $\underline{a}_0$ and proceeds to a point $\underline{a}$ of rank 1 with probability $w(\underline{a}_0, \underline{a})$; after arriving to the point $\underline{a}$, the walk continues with probability $w(\underline{a}, \underline{b})$ to a point $\underline{b}$, etc. Thus the random walk proceeds always along a directed path of G, until it reaches an endpoint, while if the path is infinite, the walk continues indefinitely. Such a random walk defines uniquely a probability measure P on the power set of the set of all paths starting from the source (this set being finite or denumerable). Let $B_{\underline{a}}$ denote the event that the random walk arrives eventually to the point $\underline{a}$. (In other words let B_a denote the set of all paths containing the point $\underline{a}$.) Let us put

$$w(a) = P(B_{\underline{a}}). \tag{2.2.}$$

Let A be any antichain, then by definition the events $B_{\underline{a}}$ $(\underline{a} \in A)$ are mutually exclusive. Thus we have for every antichain A

$$\sum_{a \in A} w(a) \leq 1 . \tag{2.3.}$$

If A is a blocking antichain, then the events $B_{\underline{a}}$ $(\underline{a} \in A)$ form a complete set of events (i.e. the sets $B_{\underline{a}}$ of paths are disjoint and their union is the set of all paths), and thus we have

$$\sum_{a \in A} w(a) = 1 . \tag{2.4.}$$

§ 3. Normal cascade graphs

We shall call a cascade graph G <u>normal</u> if the transition-probabilities $w(\underline{a}, \underline{b})$ can be chosen in such a way that $w(\underline{a})$ depends only on the rank $r(\underline{a})$ of $\underline{a}$, i.e.

(3.1.) $$w(\underline{a}) = f(r(a))$$

where f(x) is a function defined on the set of non-negative integers.

Let B_k denote the event that the random walk does not stop before arriving to a point of rank k, and let C_k denote the event that the random walk does not stop at an endpoint of rank k. Then we have evidently*

(3.2.) $$P(B_{k+1}) = P(B_k)P(C_k|B_k) .$$

Now let G be a normal cascade graph, and suppose that the transition probabilities have been chosen so that (3.1) holds: In this case we call the random walk a <u>uniform flow</u>. In case we have a uniform flow on G, clearly

(3.3.) $$P(B_k) = N_k f(k)$$

and

(3.4.) $$P(C_k|B_k) = \frac{M_k}{N_k} .$$

It follows

(3.5.) $$f(k+1) = f(k)\frac{M_k}{N_{k+1}}$$

and thus, as f(0) = 1, we get

(3.6.) $$f(k) = \frac{1}{N_k}\prod_{j<k}\frac{M_j}{N_j} \quad \text{for } k \geq 1$$

where an empty product is by definition equal to 1. Especially if the cascade graph is finite and it has no endpoits of less than maximal rank, (i.e. for every endpoint e one has r(e) = R), or if the

---

* $P(C_k|B_k)$ denotes the conditional probability of C_k under condition B_k.

cascade graph is infinite and it does not contain any endpoints, then we have $M_j = N_j$ for $j < R$ and thus

$$f(k) = \frac{1}{N_k} . \tag{3.7.}$$

The following theorem is an immediate consequence of (2.3) and (3.6):

<u>THEOREM 1.</u> <u>Let</u> G <u>be a normal cascade graph</u>. <u>Let</u> N_k <u>denote the total number of points of rank</u> k <u>of</u> G <u>and</u> M_k <u>the number of those points of rank</u> k <u>which are not endpoints</u>. <u>Let</u> A <u>be an antichain in</u> G <u>and let</u> n_k <u>denote the number of points of rank</u> k <u>in</u> A. <u>Then the inequality</u>

$$\sum_{k=0}^{R} \frac{n_k}{N_k} \prod_{j<k} \frac{M_j}{N_j} \leq 1. \tag{3.8.}$$

<u>holds</u>; <u>if</u> A <u>is a blocking antichain there is equality in</u> (3.8.).

<u>COROLLARY</u>: <u>If</u> $M_k = N_k$ <u>for</u> $k < R$, <u>where</u> R <u>is the maximal rank of points of</u> G, <u>then for every antichain</u> A <u>we have</u>

$$\sum_{k=0}^{R} \frac{n_k}{N_k} \leq 1 \tag{3.9.}$$

<u>with equality standing in</u> (3.9.) <u>if</u> A <u>is a blocking antichain.</u>

<u>Remark</u>: Notice that (3.9.) can be written also in the equivalent form

$$\sum_{a \in A} \frac{1}{N_{r(a)}} \leq 1. \tag{3.10.}$$

We shall refer to Theorem 1. as the <u>uniform flow theorem</u>. In the next § we shall give a necessary and sufficient condition for the normality of a cascade graph.

§ 4. A necessary and sufficient condition for the normality of a cascade graph

THEOREM 2. A cascade graph G is normal if and only if it satisfies the following condition: For every $k \geq 1$ and for every subset A of the set $V_k - E_k$ of points of rank k which are not endpoits, one has

$$M_k |\Gamma A| \geqq N_{k+1} |A| \tag{4.1.}$$

where N_{k+1} is the number of points of rank $k + 1$, and M_k the number of points of rank k which are not endpoints.

Proof* of Theorem 2. Clearly for every random walk (i.e. probability flow) on G and for every set of points $A \subseteq V-E$ one has

$$\sum_{\underline{a} \varepsilon A} w(\underline{a}) \leqq \sum_{\underline{b} \varepsilon \Gamma A} w(\underline{b}) \tag{4.2.}$$

Thus if $A \subset V_k - E_k$ and the flow is uniform then (4.1) holds, i.e. the condition is necessary. Let us prove now its sufficiency, i.e. that if (4.1) holds, one can choose the transition probabilities $w(\underline{a}, \underline{b})$ so that the flow should be uniform. We shall prove the existence of such transition probabilities $w(\underline{a}, \underline{b})$ step by step, i.e. by induction on the rank k of $\underline{a}$. Clearly if we put for every point $\underline{b}$ of rank 1, $w(\underline{a}_0, \underline{b}) = \frac{1}{N_1}$, then we have $w(\underline{b}) = \frac{1}{N_1}$ for every point $\underline{b}$ of rank 1. Let us suppose that we have already determined $w(\underline{a}, \underline{b})$ for all points $\underline{a}$ of rank < k so that (3.1) holds for all $\underline{a} \varepsilon V_k$. We have to show that one can choose the values of $w(\underline{a}, \underline{b})$ for all $a \varepsilon V_k$ so that (3.1) holds with $k + 1$ instead of k too.

In other words, we have to choose the transition probabilities $w(\underline{a}, \underline{b})$ for $a \varepsilon V_k - E_k$ in such a way that they should be non-negative

---

* The proof given here is due to Dr. G. Katona.

and should satisfy the following two sets of equations:

(4.3.) $$\sum_{b \in \Gamma a} w(a, b) = 1 \quad \text{for all } a \in V_k - E_k$$

and

(4.4.) $$\sum_{a \in \Gamma^{-1} b} w(a, b) = \frac{M_k}{N_{k+1}} \quad \text{for all } b \in V_{k+1}$$

Let $a_1, a_2, \ldots a_{M_k}$ and $b_1, b_2, \ldots, n_{N_{k+1}}$ denote the elements of the sets $V_k - E_k$ and V_{k+1} respectively. We consider the auxiliary graph G* defined as follows: G* has $2M_k N_{k+1}$ points which we denote by $a_{i,j}$ $(1 \leq i \leq M_k ; 1 \leq j \leq N_{k+1})$ and $b_{u,v}$ $(1 \leq u \leq N_{k+1} ; 1 \leq v \leq M_k)$. We connect the points $a_{i,j}$ and $b_{u,v}$ in G* if and only if there is in G an edge from a_i to b_u. Thus G* is a bipartite graph with the two classes of points $\mathcal{A} = \{a_{i,j}\}$ and $\mathcal{B} = \{b_{u,v}\}$. Let for any subset A* of the set $\mathcal{A}$, Γ^*A^* denote the subset of those $b_{u,v}$ which are connected by at least one $a_{i,j}$ in G*. Let A denote the set of those $a_i \in V_k - E_k$ for which $a_{i,j}$ is for at least one j contained in A*. Then we have

(4.5.) $$|A^*| \leq N_{k+1}|A|$$

and

(4.6.) $$|\Gamma^*A^*| = M_k|\Gamma A| .$$

Thus (4.1) implies

(4.7.) $$|\Gamma^*A^*| = M_k|\Gamma A| \geq N_{k+1}|A| \geq |A^*| .$$

But (4.7) means that the conditions of the <u>marriage-theorem</u> (see e.g.

Harper, L. and Rota, G. C. [4] for the existence of a one-to one matching between the sets $\mathcal{A}$ and $\mathcal{B}$ so that each $a_{i,j}$ is matched with such a $b_{u,v}$ with which it is connected in G*, are fulfilled, and thus such a matching exists. Let us take such a matching and let s(i, u) denote the number of $a_{i,j}$ $(1 \leq j \leq N_{k+1})$ which are matched to a $b_{u,v}$ $(1 \leq v \leq M_k)$. Then we have evidently

$$\sum_{u=1}^{N_{k+1}} s(i, u) = N_{k+1} \quad \text{for } 1 \leq i \leq M_k \tag{4.8.}$$

and

$$\sum_{i=1}^{M_k} s(i, u) = M_k. \tag{4.9.}$$

Further s(i, u) = 0 if a_i and b_u are not connected in G. Thus putting

$$W(a_i, b_u) = \frac{s(i, u)}{N_{k+1}} \quad (1 \leq i \leq M_k, \ b_u \in \Gamma a_i) \tag{4.10.}$$

the equations (4.3) and (4.4) hold. Thus Theorem 2 is proved.

From Theorem 2 one can easily deduce the following:

<u>COROLLARY</u>: <u>If in a cascade graph</u> G <u>the outdegree</u> D(<u>a</u>) <u>of a point</u> <u>a</u>, <u>which is not an endpoint, depends only on the rank</u> r(<u>a</u>) <u>of</u> <u>a</u>, <u>and the indegree</u> d(b) <u>of any point</u> <u>b</u> <u>depends only on the rank</u> r(b) <u>of</u> <u>b</u> <u>then</u> G <u>is normal</u>.

<u>Proof</u>: Let us denote the outdegree of a point of rank k which is not an endpoint by D_k, and the indegree of a point of rank k by d_j. As the total number of edges starting from some point of rank k is equal to the total number of edges leading to a point of rank k + 1, we have

$$M_k \cdot D_k = N_{k+1} \cdot d_{k+1} . \tag{4.11.}$$

Now let A be any subset of $V_k - E_k$. As the number of edges going out from one of the points in A cannot be larger than the number of edges arriving to a point in ΓA, we have

$$d_{k+1}|\Gamma A| \geq D_k|A| \quad . \tag{4.12.}$$

Multiplying both sides of (4.12) by N_{k+1}, and using (4.11) we get that (4.1) holds, i.e. that G is normal.

Remark. Instead of deducing the above corollary from Theorem 2 one can prove its statement by constructing effectively the uniform flow on the cascade graph. As a matter of fact, if for every a which is not an endpoint we put $w(a, b) = |\Gamma a|^{-1}$, then we get a uniform flow on the cascade graph G satisfying the conditions of the corollary.

The cascade graphs satisfying the conditions of the above Corollary of Theorem 4.1 are called semiregular cascade graphs. As by the Corollary every semiregular cascade graph is normal, it follows that the statement of Theorem 1 holds for every antichain of a semiregular cascade graph; this special case of Theorem 1 is due to Kirby A. Baker [5], who has formulated this result in a slightly different terminology, but the result expressed in our terminology is just the statement of Theorem 1 for semiregular cascade graphs.

§ 5. Kraft's inequality and Sperner's theorem as special cases of the uniform flow theorem

In spite of the extreme simplicity of its proof, Theorem 1 is a common source of several known results, thus for example Kraft's inequality (see e.g. Feinstein [6] and Sperner's theorem (see e.g. Lubell [7] . In this § we shall deduce these theorems and some of their generalizations as special instances of the uniform flow theorem.*

*I want to mention that I obtained the uniform flow theorem by analysing Lubell's elegant proof of Sperner's theorem [7].

Let us deal first with the Kraft inequality. Let us consider the cascade graph described in Example 4: Let the points of G be all finite sequences which can be formed from the digits 0, 1, ..., q-1 ($q \geq 2$) and let us draw a directed edge from the sequence $\underline{a}$ to the sequence $\underline{b}$ if and only if $\underline{b}$ is obtained by adding one more digit to the end of $\underline{a}$. Let the source be the empty sequence. It is easy to see that the cascade graph thus obtained is a tree, moreover a regular search tree in which there are q edges going out from every point and no endpoints. Let A be an antichain in G; thus A is a finite or denumerable family of finite sequences formed from the digits 0, 1, ..., q-1 such that no sequence in A is an initial segment (prefix) of another sequence in A. Such families of sequences are called in information theory q-ary prefix codes, and its elements codewords. Evidently G is a normal cascade graph; to show this we do not need Theorem 2, because it is easy to see that putting $w(\underline{a}, \underline{b}) = q^{-1}$ for every edge $\underline{a}\,\underline{b}$ of G we get $w(\underline{a}) = q^{-r(\underline{a})}$, i.e. a uniform flow. Applying Theorem 1 we obtain the following result, called Kraft's inequality:

If A is a q-ary prefix code ($q \geq 2$) and n_k denotes the number of codewords of length k in the code, then the inequality

$$\sum_{k=0}^{\infty} \frac{n_k}{q^k} \leq 1 \tag{5.1.}$$

holds; if A has the property that every infinite sequence consisting of the digits 0, 1, ..., q-1 contains one of the codewords from A as an initial segment, then there is equality in (5.1).

Clearly (5.1) can be formulated also in another form, speaking about trees instead of codes. It is natural to ask in general which rooted trees, considered as cascade graphs, are normal. The answer is very simple: A rooted tree is a normal cascade graph if and only if it is semiregular, i.e. the outdegree of any point which is not an endpoint depends only on the rank of the point. If $D(a) = q_k$ for all $a \in V_k - E_k$ then $N_{k+1} = M_k \cdot q_k$ and thus we get from the uniform flow theorem the following result, which reduces to the Kraft inequality in the special case when $q_k = q$ for all k.

If A is an antichain in a semiregular rooted tree such that $D(a) = q_k$ if $a \in V_k - E_k$ and if n_k denotes the number of elements of the antichain A having rank r, then the inequality

$$(5.2.)\qquad \sum_{k=0}^{\infty} \frac{n_k}{q_1 q_2 \cdots q_k} \leq 1$$

holds.

Of course Kraft's inequality (as well as its generalization) can be proved quite easily directly, but it is instructive to consider this inequality as a particular instance of the uniform flow theorem.

Now let us consider how Sperner's theorem is obtained from Theorem 1. Let G be the cascade graph of Example 2 of § 1. Let the points of G be all subsets of an n-element set S, and let there be an edge in G from the set $\underline{a} \subseteq S$ to the set $\underline{b} \subseteq S$ if and only if $\underline{b}$ is obtained from $\underline{a}$ by omitting one of its elements. To show that the cascade graph G is normal it is sufficient to point out that if we put

$$(5.3.)\qquad w(\underline{a}, \underline{b}) = \frac{1}{n - r(\underline{a})} \quad \text{for every } \underline{a} \text{ with } r(\underline{a}) < n$$

(notice that $r(a) = n$ only if $\underline{a}$ is the empty set and this is the (unique) endpoint of G), then we get, taking into account that there are $(n-k)!$ paths from the source to a point a for which $r(\underline{a}) = n - k$ i.e. to set $\underline{a}$ having k elements, it follows that

(5.4.) $$w(\underline{a}) = \frac{1}{\binom{n}{r(\underline{a})}}$$ for all points a,

i.e. there exists a uniform flow in G. (This follows also from the corollary of Theorem 2. It is easy to see further that a subset A of the points of G, i.e. a family of subsets of S in an antichain if for any two different sets $\underline{a} \subseteq S$ and $\underline{b} \subseteq S$ belonging to A, $\underline{a}$ is not a subset of $\underline{b}$, i.e. if A is a Sperner-system of subsets S. Thus the iform flow theorem yields the following result:

If A is a Sperner system of subsets of an n-element set, and A contains n_k sets having k elements (k=0,1,...) then the inequality

(5.5.) $$\sum_{k=0}^{n} \frac{n_k}{\binom{n}{k}} \leq 1$$

holds.

As

(5.6.) $$\max_k \binom{n}{k} = \binom{n}{[\frac{n}{2}]},$$

we obtain from (5.6) the usual (though slightly weaker) form of Sperner's theorem:

(5.7.) $$|A| \leq \binom{n}{[\frac{n}{2}]}.$$

Before going further let us add a remark. Comparing the two special cases just discussed, it turns out that the property of a system of sets being a Sperner-system plays the same role in Sperner's theorem as the prefix property of a code in Kraft's inequality. As a matter of fact, there is a real connection between these two concepts, not only a superficial analogy.

As mentioned earlier the prefix property of a code implies that if the codewords of such a code are written one after the other, without indicating where one codeword ends and the next begins, and if the code has the prefix property, the sequence of symbols can be uniquely decoded, i.e. the codewords can be unambiguously separated from another. Now let us impose on the code the auxiliary restriction that a sequence of codewords should be uniquely decodable even in the case when the letters within each codeword are arbitrarily rearranged, i.e. if the codewords are unordered sets of letters, and not ordered sets as usual. If we require further that the same letter should not occur more than once in any codeword, then such a code is uniquely decodable if the codewords (considered as unordered sets of letters) form a Sperner-system. Expressed in the language of search theory*, a Sperner-system corresponds to such a strategy of search in which the values of the test functions are obtained simultaneously and one does not know which value comes from which test function.

For example let us consider the 19 numbers

0,1,2,3,4,5,6,7,8,12,13,14,16,17,18,19,23,24,29.

These numbers can be uniquely characterized by their residues mod 2, mod 3, mod 5, even if these three residues are given in a random order. For instance if we are told that the three residues are 0, 0, and 2, it is easy to see that among the 19 numbers enlisted above only 20 has these residues, namely it is congruent to 0 mod 2 and mod 5 and to 2 mod 3.

As a further application of the uniform flow theorem, we prove the following generalization of Sperner's theorem:

THEOREM 3. Let $\mathcal{A}$ be a family of ordered r-tuples ($r \geq 1$)

---

* As regards the connection of the results of this paper with search theory, see [8].

of disjoint subsets of an n-element set S, such that if $(A_1, A_2, \ldots, A_r)$ and $(B_1, B_2, \ldots, B_r)$ both belong to the family $\mathcal{A}$ then the relations $A_j \subseteq B_j$ $(j=1,1,\ldots,r)$ cannot hold simultaneously. Then the number of elements of the family $\mathcal{A}$ satisfies the inequality

$$(5.8.)\qquad |\mathcal{A}| \leq r^{\left[\frac{r(n+1)}{r+1}\right]} \cdot \binom{n}{\left[\frac{r(n+1)}{(r+1)}\right]}.$$

which is best possible, i.e. equality is possible in (5.8) by a suitable choice of the class $\mathcal{A}$.

Proof of Theorem 3. Let us construct a cascade graph G as follows: The points of G are all possible $(r + 1)^n$ ordered r-tuples $(A_1, A_2, \ldots, A_r)$ of disjoint subsets of S. The source of G is the r-tuple each element of which is the empty set. The rank of an r-tuple $\underline{a} = (A_1, A_2, \ldots, A_r)$ is $r(\underline{a}) = \sum_{j=1}^{r} |A_j|$ and there is an edge from $\underline{a} = (A_1, A_2, \ldots, A_r)$ to $\underline{b} = (B_1, B_2, \ldots, B_r)$ if and only if the following conditions are satisfied: $A_j \subseteq B_j$ for $j=1,2,\ldots,r$ and $r(\underline{b}) = r(\underline{a}) + 1$; in other words there is in G an edge from $\underline{a}$ to $\underline{b}$ if $A_j = B_j$ for all but one value of j $(1 \leq j \leq r)$--say except for i --and B_i is obtained from A_i by adding one more element of S to A_i. Clearly the conditions of Theorem 3 mean that $\mathcal{A}$ should be an antichain in G. Thus if we show that G is normal, we can apply Theorem 1.

We shall prove the normality of G by verifying that the conditions of the corollary to Theorem 1 are fulfilled. Let $a = (A_1, A_2, \ldots, A_r)$ be any point in G which is not an endpoint; then we have $r(\underline{a}) < n$ because the endpoints of G are the points with $r(\underline{a}) = n$. Thus there are $r - r(\underline{a})$ elements of S which can be added to one of the sets

A_1 to get a point $\underline{b}$ to which there leads an edge from $\underline{a}$: The number of possible choices is thus $r\cdot(n-r(\underline{a}))$. Thus $D(a) = r\,(n-r(\underline{a}))$ depends only on the rank of $\underline{a} \in V - E$. Let now $\underline{b}$ be any point of G. By a similar argument we get that $d(\underline{b}) = r(\underline{b})$. Thus all conditions of the Corollary of Theorem 1 are satisfied, and therefore G is normal. Thus we can apply Theorem 1 to the antichain $\mathcal{A}$ and we get, in view of $N_k = r^k \binom{n}{k}$ and $M_k = N_k$ if $k < n$, that

(5.9.)

$$\sum_{k=0}^{n} \frac{n_k}{r^k \binom{n}{k}} \le 1,$$

where n_k denotes the number of elements of $\mathcal{A}$ having rank k.
As however

$$(5.10) \qquad \max_{0 \leq k < n} r^k \binom{n}{k} = r^{\left[\frac{(n+1)r}{r+1}\right]} \binom{n}{\left[\frac{(n+1)r}{(r+1)}\right]}, \text{ and } \sum_{k=0}^{n} n_k = |\mathcal{A}|$$

it follows that (5.8) holds.

Clearly for r = 1 Theorem 3 is nothing else than Sperner's theorem. To show that the inequality (5.8) is best possible take for $\mathcal{A}$ the family of all r-tuples $(A_1, A_2, \ldots, A_r)$ of disjoint subsets of S such that $\sum_{j=1}^{r} |A_j| = \left[\frac{(n+1)r}{r+1}\right]$ (i.e. all points of rank $\left[\frac{(n+1)r}{(r+1)}\right]$ of G); $\mathcal{A}$ clearly satisfies the requirements of the Theorem 3 and $|\mathcal{A}|$ is equal to the right hand side of (5.8).

Other generalizations of Sperner's theorem can also be obtained from Theorem 3. We intend to return to these elsewhere.

June 18, 1969

References

1 Henry H. Crapo and Gian-Carlo Rota, On the foundation of Combinatorial theory: Combinatorial Geometries, M.I.T. 1968.

2 G. Birkhoff, Lattice Theory, Amer. Math. Soc. Coll. Publ.

3 David A. Klarner, The number of graded partially ordered sets, Journal of Combinatorial Theory 6 (1969) p. 12-19.

4 L. Harper and G. C. Rota, Matching theory; an introduction. M.I.T.

5 Kirby A. Baker, A generalization of Sperner's Lemma, Journal of Comninatorial Theory 6 (1969) p. 224-225.

6 A. Feinstein, Foundations of information theory, McGraw - Hill, 1958.

7 D. Lubell, A short proof of Sperner's Lemma, Journal of Combinatorial Theory 1 (1966) p. 299.

8 A. Rényi, Lectures on the theory of search, University of North Carolina at Chapel Hill, Institute of Statistics Mimeo Series No. 600.7, May 1969.

AN APPROACH TO THE THEORY OF PRESSURE BROADENING OF SPECTRAL LINES

Wilhelm von Waldenfels *)**)

A b s t r a c t

An attempt has been made to give a rigorous treatment of pressure or Stark broadening under the following assumptions:
1) the perturbating particles move independent of each other on straight lines with constant velocity v. The particles are distributed in the space with mean density n, the directions of their velocities are uniformly distributed on the unit sphere.
2) We neglect at all interactions of particles whose impact parameter is bigger than some constant ρ and we neglect interactions of particles with impact time t_p at times t with $|t - t_p| \geq \tau/2$ where τ is another constant. Under these assumptions we get two formulae for the line profile $I(\omega)$ or its Fourier transform $R(t) = \int I(\omega)e^{i\omega t}d$, Part II, Theorem 3 and 4. Both formulae are very similar in character. They consist of terms which are sums of one-particle-interactions, two-particle-interactions and so on. The first formula holds for any value of n, v, τ and ρ whereas the second one can only be proven for $n\pi\rho^2 v\tau < \log 2$. But the second formula is independent of τ and its one-particle approximation even of ρ. A detailed discussion of the second formula has been made in I. It

*) Institut für Angewandte Mathematik der Universität Heidelberg, W. Germany.

**) The research reported herein has been sponsored in part by the United States Government.

clearly shows the nature of impact approximation.

For $\omega^2 I(\omega)$ there can be established another very similar type of formulae (cf. II.5.). This type seems to be very useful for the discussion of the line-wings (cf. the end of I.).

The mathematical techniques are mainly the well-known methods of nearly elementary probability theory. New seems to me the calculus of many-particle-interactions developped in II.2. and applied in II.3.

Table of contents

I. Discussion of problem and results

The general formula for the shape of a spectral line emitted by an atom or ion under the influence of a stochastic perturbation is essentially given by

$$J(\omega) = \frac{1}{2\pi} E \int e^{-i\omega t} \, \mathrm{Tr}\mu(0)\mu(t)dt$$

where $\mu(t)$ is the dipole momentum at time t and E signifies the statistical expectation.

We have*)

$$\mu(t) = U(t)^*\mu(0)U(t)$$

where $U(t)$ is the time development operator given by the differential equation

$$\frac{d}{dt} U(t) = \frac{i}{\hbar} H(t)U(t)$$

$$U(0) = 1$$

and $H(t)$ is the Hamiltonian of the system.

We introduce the notation of the product integral ϖ (the letter ϖ is the calographic version of π): if we have a matrix function $X(t)$ with $X(t) = A(t) X(t)$ and $X(t_o) = 1$, then we write

$$X(t) = \varpi_{t_o}^{t} (1 + A(s)ds).$$

If A is Hermitian, then $\varpi_{t_o}^{t} (1 + iA(s)ds)$ is unitary and we write

---

*) We use throughout this paper three types of conjugates:

1) the conjugate complex denoted by $\overline{X}$,
2) the Hermitian conjugate matrix denoted by X^*,
3) the Hermitian conjugate and time reversed matrix function $x^+ : x^+(t) = (x(-t))^*$.

for short

$$\underset{t_o}{\overset{t}{\tilde{\omega}}} (1 + iA(s)ds) = \underset{t_o}{\overset{t}{\tilde{\omega}}} A.$$

So

$$U(t) = \underset{o}{\overset{t}{\tilde{\omega}}} H/\hbar.$$

We are interested in a line caused by a transition of a level a to another level b. Let $\mathfrak{H}_a$ and $\mathfrak{H}_b$ be the finite dimensional Hilbert spaces spanned by the eigenstates of level a, resp. level b. The spaces $\mathfrak{H}_a$ and $\mathfrak{H}_b$ are orthogonal. All operators have to be interpreted as operators on $\mathfrak{H}_a \oplus \mathfrak{H}_b$, the trace Tr being defined accordingly. The operator $\mu(0)$ is of the form

$$\mu(0) = \begin{pmatrix} 0 & \mu_{ab} \\ \mu_{ba} & 0 \end{pmatrix} .$$

where μ_{ab} is an operator from $\mathfrak{H}_b$ to $\mathfrak{H}_a$ and $\mu^*_{ba} = \mu_{ab}$. The Hamiltonian splits into

$$H(t) = H_o + V(t)$$

with

$$H_o = \begin{pmatrix} E_a & 0 \\ 0 & E_b \end{pmatrix}.$$

We assume that the stochastic perturbation $V(t)$ does not cause any transitions from level a to b or from b to a. Therefore

$$V(t) = \begin{pmatrix} V_a(t) & 0 \\ 0 & V_b(t) \end{pmatrix}$$

and

$$U(t) = \begin{pmatrix} e^{+iE_a t/\hbar}\, T_a(t) & 0 \\ 0 & e^{iE_b t/\hbar}\, T_b(t) \end{pmatrix}$$

with

$$T_a(t) = \mathop{\varpi}_{0}^{t} V_a/\hbar$$

$$T_b(t) = \mathop{\varpi}_{0}^{t} V_b/\hbar .$$

One gets

$$\mathrm{Tr}\,\mu(0)\,\mu(t) = e^{-i\omega_o t}\,\mathrm{Tr}\,\mu_{ba}T_a^{*}\mu_{ab}T_b$$

$$+\, e^{+i\omega_o t}\,\mathrm{Tr}\,\mu_{ab}T_b^{*}\mu_{ba}T_a$$

with

$$\omega_o = \frac{E_a - E_b}{\hbar} .$$

Writing the matrices with all indices one obtains

$$\mathrm{Tr}\mu_{ab}\ T_b^{*}\ \mu_{ba}\ T_a =$$

$$= \sum_{\alpha',\alpha'',\beta',\beta''} (\mu_{ab})_{\alpha'\beta'}(T_b^{*})_{\beta'\beta''}(\mu_{ba})_{\beta''\alpha''}(T_a)_{\alpha''\alpha'}$$

$$= \sum_{\alpha',\alpha'',\beta',\beta''} (\mu_{ba})_{\beta''\alpha''}(T_a)_{\alpha''\alpha'}(\overline{T}_b)_{\beta''\beta'}(\mu_{ab})_{\alpha'\beta'} .$$

The matrix

$$(T_a)_{\alpha''\alpha'}\ (\overline{T}_b)_{\beta''\beta'}$$

can be interpreted as the matrix of the unitary operator $T_a \otimes \overline{T}_b$ on the tensor product $\mathfrak{H}_a \otimes \mathfrak{H}_b$. We get

$$T_a \otimes \overline{T}_b = \mathop{\varpi}_{0}^{t} (V_a \otimes 1_b - 1_a \otimes \overline{V}_b)/\hbar$$

where 1_a and 1_b are the identity operators on $\mathfrak{H}_a$ resp. $\mathfrak{H}_b$. Call

$$I(\omega) = \frac{1}{2\pi}\, E \int e^{-i\omega t}\, dt\ T_a(t) \otimes \overline{T}_b(t) ,$$

then the line profile is

$$J(\omega) = \sum_{\alpha',\alpha'',\beta',\beta''} (\mu_{ba})_{\beta''\alpha''} \Big(I(\omega - \omega_o)_{\alpha''\beta'',\alpha'\beta'} +$$

$$+ I(\omega + \omega_o)_{\alpha'\beta',\alpha''\beta''} \Big) (\mu_{ab})_{\alpha'\beta'}.$$

So all is reduced to the calculation of $I(\omega)$.

The general mathematical frame is the following: given a stochastic stationary process $X(t)$, where $X(t)$ is a Hermitian operator on a Hilbert space $\mathfrak{h}$, calculate

(1) $$I(\omega) = \frac{1}{2\pi} \int e^{-i\omega t} \, dt \, E \overset{t}{\underset{0}{\tilde{\omega}}} X.$$

In our case $\mathfrak{h} = \mathfrak{h}_a \otimes \mathfrak{h}_b$ and

$$X(t) = (V_a(t) \otimes 1_b - 1_a \otimes \overline{V}_b(t))/\hbar.$$

Stationarity of the process X means physically that we have a physical stationary situation, e.g. the atom is placed into an ideal gas of constant temperature and density.

For every ω is $I(\omega)$ a positive definite matrix and

$$\int \mathrm{Tr}\, I(\omega) \, d\omega = 1.$$

The last equation follows out of the fact that

$$R(t) = \int I(\omega) \, e^{i\omega t} \, dt = E \overset{t}{\underset{0}{\tilde{\omega}}} X$$

and that $\mathrm{Tr}\, R(0) = 1$. The fact that $I(\omega)$ is a positive matrix can directly be seen from the equation

(2) $$I(\omega) = \lim_{T \mapsto \infty} \frac{1}{2\pi T} E \left(\int_0^T e^{-i\omega t} \overset{t}{\underset{0}{\tilde{\omega}}} X \, dt \right) \left(\int_0^T e^{-i\omega t} \overset{t}{\underset{0}{\tilde{\omega}}} X \, dt \right)^*.$$

This well-known formula (cf. [3]) can easily be derived.

The product integral has the properties

$$\overset{c}{\underset{b}{\tilde{\omega}}} \; \overset{b}{\underset{a}{\tilde{\omega}}} = \overset{c}{\underset{a}{\tilde{\omega}}}$$

especially

$$\begin{pmatrix} b \\ \tilde{\omega} \\ a \end{pmatrix}^{-1} = \overset{a}{\underset{b}{\tilde{\omega}}} .$$

As $\overset{s}{\underset{0}{\tilde{\omega}}} X$ is unitary, we have

$$\left(\overset{s}{\underset{0}{\tilde{\omega}}} X\right)^{*} = \left(\overset{s}{\underset{0}{\tilde{\omega}}} X\right)^{-1} = \overset{0}{\underset{s}{\tilde{\omega}}} X$$

and

$$\left(\overset{t}{\underset{0}{\tilde{\omega}}} X\right)\left(\overset{s}{\underset{0}{\tilde{\omega}}} X\right)^{*} = \overset{t}{\underset{s}{\tilde{\omega}}} X.$$

The process X is assumed to be stationary, therefore

$$E \overset{t}{\underset{s}{\tilde{\omega}}} X = E \overset{t-s}{\underset{0}{\tilde{\omega}}} X = R(t-s).$$

The Fourier transform of the expression behind the limes sign is equal to

$$\frac{1}{2\pi T} \int e^{i\omega\xi} \, d\omega \int_0^T \int_0^T e^{-i\omega(t-s)} R(t-s) dt ds =$$

$$= \begin{cases} 0, & \text{if } |\xi| > T \\ (1 - |\xi|/T) \, R(\xi), & \text{if } |\xi| < T \end{cases}$$

$$\longrightarrow R(\xi) \quad \text{for} \quad T \longrightarrow \infty .$$

This proves equation (2).

We assume now that the light emitting atom is placed into an ideal gas consisting of particles moving on straight lines with constant velocity v. The particles are independent of each other, their mean density is n, the directions of their velocities are uniformly distributed on the unit sphere.

Let the light emitting atom at rest at the origin. Let $\vec{p}_\iota$ be the coordinate vector of the point of nearest approach of the ι-th particle to the origin and let t_ι be the corresponding time. Then the coordinate of the ι-th particle is given by

$$\vec{x}_\iota(t) = \vec{p}_\iota + v\vec{e}_\iota(t - t_\iota)$$

where $\vec{e}_\iota$ is the direction of velocity. We call $\vec{p}_\iota$ the vectorial impact parameter and t_ι the impact time of the ι-th particle.

Now we want to investigate more in detail the nature of the quantity

$$X(t) = (V_a \otimes 1_b - 1_a \otimes \bar{V}_b)/\hbar.$$

Consider, for instance, the case of a hydrogen atom placed into a gas of electrons. In this case the perturbating operator $V(t) = e\vec{r}\cdot\vec{E}(t)$ where $\vec{r}$ is the operator of the radius vector and $\vec{E}(t)$ is the electrostatic force due to the electrons. Denoting by $\vec{r}_a$ the matrix $(\vec{r})_{\alpha'\alpha''}$ where α',α'' run through an orthonormal basis of eigenstates of level a, we get

$$\begin{aligned} X(t) &= (\vec{r}_a \otimes 1_b - 1_a \otimes \bar{\vec{r}}_b)/\hbar \cdot \vec{E}(t) \\ &= \vec{A}\cdot\vec{E}(t) \end{aligned}$$

where $\vec{A}$ is a constant matrix. We have

$$\begin{aligned} \vec{E}(t) &= \sum_\iota e\vec{x}_\iota(t)\,|\vec{x}_\iota(t)|^{-3} \\ &= \sum_\iota \vec{E}_\iota(t) \end{aligned}$$

where $\vec{E}_\iota(t)$ is the field strength due to the ι-th particle. So

$$X(t) = \vec{A}\cdot\sum_\iota \vec{E}_\iota(t).$$

Consider now the effect of the electron gas on a line with quadratic Stark effect. Then $X(t)$ is a real number, it is the instantaneous frequency perturbation and

$$\begin{aligned} X(t) &= \text{const}\ |\vec{E}(t)|^2 \\ &= \text{const}\ |\textstyle\sum_\iota \vec{E}_\iota(t)|^2. \end{aligned}$$

Therefore we assume the following general form for $X(t)$

(3) $$X(t) = h(\sum_{\iota} \varphi_{\iota}(t))$$

where h is continuous and $h(0) = 0$. In the preceding two examples there was $\varphi_{\iota}(t) = \vec{E}_{\iota}(t)$, in the first example h is the function $\vec{\xi} \longmapsto \vec{A} \cdot \vec{\xi}$, in the second example h is $\vec{\xi} \longmapsto \text{const} |\vec{\xi}|^2$.
The function $\varphi_{\iota}(t)$ depends on $\vec{p}_{\iota}$, t_{ι} and some interior parameters of the ι-th particle, e.g. the dipole momentum and its angular velocities. We give the new symbol ζ_{ι} to $\vec{p}_{\iota}$ and to the set of interior parameters together. Then

(4) $$\varphi_{\iota}(t) = \varphi(\zeta_{\iota}, t - t_{\iota}).$$

Throughout the paper we make the following assumptions which seem somewhat reasonable:

$$\text{(i)} \quad \varphi(\zeta, t) = 0 \quad \text{for} \quad |t| > \tau/2$$

$$\text{(ii)} \quad \varphi(\zeta, t) = 0 \quad \text{for} \quad |\vec{p}| > \rho$$

where $\tau > 0$ and $\rho > 0$ are some arbitrary constants which should be chosen according to the physical situation. Both assumptions together represent a cut-off outside a cylinder of radius ρ and length τv, the direction of the cylinder is the direction of the perturbating particle, the center of the cylinder is the origin. One often uses cut-offs outside a sphere, but it makes no great difference to replace a sphere by a cylinder. The cylinder has some mathematical advantages, because the time of interaction of every particle is equal to τ and does not depend on $\vec{p}$.
The second assumption implies that we have not to count the particles ι with impact parameter $|p_{\iota}| > \rho$. Call t_1 the first impact time of a particle with impact parameter $\leqslant \rho$ after $t_o = 0$, call t_2

the second time. Then

$$t_o < t_1 < t_2 < \dots .$$

The independence of the particles and their uniform distribution yields that the differences

$$u_k = t_{k+1} - t_k, \quad k = 0, 1, 2, \dots$$

are independent and that they are identically distributed with respect to the law

$$\text{Prob}\{u \in x, x + dx\} = ce^{-cx}\,dx$$

with $c = vn\pi\rho^2$.

The parameters $\zeta_1, \zeta_2, \dots$ corresponding to $t_1, t_2, \dots$ are independent and identically distributed, too. If, e.g., the perturbating particle is completcly described by $\zeta = (\vec{p}, \vec{e})$, then ζ varies on

$$Z = \{(\vec{p}, \vec{e}) \in \mathbb{R}^6 : \vec{p}\cdot\vec{e} = 0, \ |\vec{e}| = 1, |\vec{p}| \leqslant \rho\}$$

and the distribution of ζ is given by the formula

$$Ef(\zeta) = \frac{1}{\pi\rho^2} \int_S d^2\vec{e} \int_{\substack{\vec{p}\cdot\vec{e} = 0 \\ |\vec{p}| \leqslant \rho}} d^2\vec{p}\, f(\vec{p}, \vec{e})$$

where $d^2\vec{e}$ is the normalized Lebesgue measure on the unit sphere S, i.e. $\int_S d^2e = 1$ and $d^2\vec{p}$ is the Lebesgue measure on the disc $\vec{p} \perp \vec{e}, |\vec{p}| \leqslant \rho$.

We replace (2) by (5)

$$(5) \qquad I(\omega) = \lim_{N \longrightarrow \infty} \frac{c}{2\pi N} E\left(\int e^{-i\omega t} W_N(t)dt\right)\left(\int e^{-i\omega t} W_N(t)dt\right)^*$$

$$W_N(t) = \overset{t}{\underset{-\infty}{\tilde{\omega}}} \Phi_N - \mathbf{1}_{\leqslant o}(t) - \mathbf{1}_{>u_1+\ldots+u_N}(t) \overset{+\infty}{\underset{-\infty}{\tilde{\omega}}} \Phi_N$$

$$\Phi_N = h(\varphi(\zeta_1) + \varphi(\zeta_2)[u_1] + \ldots +' \varphi(\zeta_N)[u_1 + \ldots + u_{N-1}]).$$

Here $\mathbf{1}_{\leqslant}(t)$ is equal to 1 for $t \leqslant 0$ and equal to 0 elsewhere. The $[\ldots.]$ denote the translation. So $\varphi(\zeta_3)[u_1 + u_3]$ is the function $t \longmapsto \varphi(\zeta_3, t - (u_1 + u_2))$.

The main reason for the validity of (5) is that $\Phi_N(t) = X(t - t_1)$ for $\tau/2 \leqslant t \leqslant u_1 + \ldots + u_{N-1} - \tau/2$ and that for large N the stochastic variable $u_1 + \ldots + u_{N-1}$ behaves like N/c. So one replaces in (2) the integrand by W_N, which vanishes outside $[-\tau/2, u_1 + \ldots + u_N + \tau/2]$ and then T in the dominator by N/c. We shall not prove formula (5) in detail. One has to perform a Fourier transform and then some lengthy but straightforward calculations.

Equation (5) is the starting point of our mathematical treatment. For physics the most important formula coming out of this treatment is

$$(6) \qquad I(\omega) = \tilde{\tilde{B}} + (1 + cB) \frac{1}{i\omega + c - cA} (1 + c\tilde{B}) +$$
$$+ (1 + c\tilde{B}^*) \frac{1}{-i\omega + c - cA^*} (1 + cB^*)$$

with

$$A(\omega) = A_1(\omega) + A_2(\omega) + \ldots$$

$$B(\omega) = B_1(\omega) + B_2(\omega) + \ldots$$

$$\tilde{B}(\omega) = \tilde{B}_1(\omega) + \tilde{B}_2(\omega) + \ldots$$

$$\tilde{\tilde{B}}(\omega) = \tilde{\tilde{B}}_1(\omega) + \tilde{\tilde{B}}_2(\omega) + \ldots$$

These series converge for $c\tau < \log 2$. They are not directly power series but they proceed by powers of $c\tau$, i.e. $A_2(\omega)$ is by a factor $\approx c\tau$ less than A_1, and A_3 by a factor $\approx c\tau$ less than A_2, and so on. A_1 contains the one-particle perturbation, A_2 the two-particle perturbations, ... The same statements hold for the other series. I do not know whether the condition $c\tau < \log 2$ has physical importance or not. In the moment I am only able to prove the convergence in this case, but my proof of convergence uses very crude estimates.

The important quantity is $c\tau = \pi\rho^2 n v\tau$. It seems reasonable to set $v\tau$ and ρ equal to the Weisskopf radius ρ_W (cf. [3]). Therefore $c\tau = \pi n\rho_W^3$. This is the mean number of particles in a cylinder of length and radius equal to ρ_W. It is well-known as a critical parameter from other papers ([1], [3]).

In the second part of the paper we show another similar formula (theorem 2), which has the advantage of convergence for any value of $c\tau$. The big disadvantage of this formula is, however, that it depends heavily on the choice of τ, which is a somewhat artificial quantity. Formula (6) does not depend on τ, in the case of quadratic Stark effect its first order approximation does not even depend on the other cutting-off-parameter ρ. So it might be a good idea, especially for low densities, i.e. $n\rho_W^3 \ll 1$, to compare the first order approximation of (6) with the experiment.

Call

$$A(1) = \mathop{\bar{\omega}}_{-\infty}^{+\infty} h(\varphi(\zeta))$$

$$B(1) = \int_0^{\infty} e^{-i\omega t}\, dt \left(\mathop{\bar{\omega}}_{-\infty}^{t} h(\varphi(\zeta)) - \mathop{\bar{\omega}}_{-\infty}^{+\infty} h(\varphi(\zeta)) \right) + \int_{-\infty}^{0} e^{-i\omega t}\, dt \left(\mathop{\bar{\omega}}_{-\infty}^{t} h(\varphi(\zeta)) - 1 \right)$$

then

$$A_1 = E\,A(1)$$
$$B_1 = E\,B(1)$$
$$\tilde{B}_1 = E\,A(1)B(1)^*$$
$$\tilde{\tilde{B}}_1 = E\,B(1)B(1)^*.$$

The higher terms are a bit more complicated. There exists a general formula, which is completely described in Chapter II. As an example we note

$$A_2 = E\,e^{-i\omega u_1}\left(\overset{+\infty}{\underset{-\infty}{\varpi}} h(\varphi(\zeta_1) + \varphi(\zeta_2)[u_1]) - \overset{+\infty}{\underset{-\infty}{\varpi}} h(\varphi(\zeta_2)) \overset{+\infty}{\underset{-\infty}{\varpi}} h(\varphi(\zeta_1))\right).$$

The expressions ζ, ζ_1, ζ_2, u_1 are stochastic variables which obey the laws described above.

Consider broadening by quadratic Stark effect. Then $\zeta = (\vec{p}, \vec{e})$ and

$$\varphi(\vec{p}, \vec{e})(t) = -e(\vec{p} + v\vec{e}t)|\vec{p} + v\vec{e}t|^{-3}$$
$$h(\vec{X}) = -2\pi C/e \cdot |\vec{X}|^2.$$

In this formula we dropped already the cutting-off-parameter τ. In order to avoid ρ, consider

$$a = c(A_1 - 1) = nv\pi\rho^2\left(\frac{1}{\pi\rho^2}\int_S d^2\vec{e} \int_{\substack{\vec{p}\perp\vec{e} \\ |\vec{p}|\leqslant\rho}} d^2\vec{p}\right.$$
$$\left.\left(\exp i\int_{-\infty}^{+\infty} 2\pi\alpha(p^2 + v^2t^2)^{-2}\,dt - 1\right)\right).$$

In this formula we can go without difficulty with ρ to infinity and get

$$a = nv\int_0^{+\infty} 2\pi p\,dp\left(\exp i\int_{-\infty}^{+\infty} 2\pi C(p^2 + v^2t^2)^{-2}\,dt - 1\right).$$

In the same way one obtains with

$$\Psi(t, p) = \exp i \int_{-\infty}^{t} 2\pi C(p^2 + v^2 s^2)^{-2} \, ds$$

$$b = cB_1 = c\tilde{B}_1 = nv \cdot \int_0^{\infty} 2\pi p dp.$$

$$\left[\int_0^{\infty} e^{-i\omega t} dt(\Psi(t, p) - \Psi(\infty, p)) + \int_{-\infty}^{0} e^{-i\omega t} dt(\Psi(t, p) - 1)\right]$$

$$\tilde{\tilde{b}} = c\tilde{\tilde{B}}_1 = nv \int_0^{\infty} 2\pi p dp$$

$$\left|\int_0^{\infty} e^{-i\omega t} dt(\Psi(t, p) - \Psi(\infty, p)) + \int_{-\infty}^{0} e^{-i\omega t} dt(\Psi(t, p) - 1)\right|^2$$

and finally

$$(7) \qquad 2\pi I(\omega) = \tilde{\tilde{b}} + \frac{(1 + b)^2}{i\omega - a} + \frac{(1 + \bar{b})^2}{-i\omega - \bar{a}} .$$

If we consider broadening by linear Stark effect, then φ is the same as above but $h(x) = \vec{A}\cdot\vec{x}$, where $\vec{A}$ is a vector whose components are Hermitian matrices. Well-known convergence difficulties oblige in this case to retain a cutting-off-parameter ρ, which is usually equal to the Debye radius or to the interatomic distance. No big improvement of formula (6) is possible in this case. It should, however, be pointed out that, as far as we know, this formula is the first exact formula for broadening by linear Stark effect.

We conclude by a discussion of the first order formula considering only one-particle-interactions under the assumption $c\tau \ll 1$.

$$(8) \quad 2\pi I(\omega) = c\tilde{\tilde{B}}_1 + (1 + cB_1)(i\omega + c - cA)^{-1}(1 + c\tilde{B}_1) + $$
$$+ (1 + c\tilde{B}_1^*)(-i\omega + c - cA^*)^{-1}(1 + cB_1^*).$$

We have two different characteristic frequencies c and $1/\tau$. Let us at first assume that $|\omega| \approx c$. The quantities cB_1 and $c\tilde{B}_1$ are of order $c\tau$, the quantity $c\tilde{\tilde{B}}_1$ is of order $c\tau^2$. We can neglect them and obtain

$$2\pi I(\omega) = \frac{1}{i\omega + c - cA_1} + \frac{1}{-i\omega + c - cA_1^*}\,, \quad |\omega| \lessapprox c\,.$$

This is the well-known impact approximation (cf. [3], [2], p. 69).

In the other frequency range $|\omega| \approx \frac{1}{\tau} >> c$, we have at first to switch to another formula. Put

$$\Gamma(1) = \int e^{-i\omega t}\, h(\varphi(\zeta))(t) \overset{t}{\underset{-\infty}{\tilde{\omega}}} h(\varphi(\zeta))dt$$

i.e. the Fourier transform of the derivative of $\overset{t}{\underset{-\infty}{\tilde{\omega}}} h(\varphi(\zeta))$, and

$$\Gamma_1 = E\Gamma(1), \quad \tilde{\Gamma}_1 = EA(1)\Gamma(1)^*, \quad \tilde{\tilde{\Gamma}}_1 = E\Gamma(1)\Gamma(1)^*.$$

Then

$$(9) \qquad 2\pi\omega^2 I(\omega) = c\tilde{\tilde{\Gamma}}_1 + c\Gamma_1 \frac{1}{i\omega + c - cA}\, c\tilde{\Gamma}_1 + c\tilde{\Gamma}_1^* \frac{1}{i\omega + c - cA}\, c\Gamma_1.$$

If M is the maximum value of the norm of $h(\varphi(\zeta))$, then $c\tilde{\tilde{\Gamma}}_1$ is of the order $cM^2\tau^2$ and $c\Gamma$ and $c\tilde{\Gamma}_1$ of the order $Mc\tau$. If $\omega \approx 1/\tau$, then the second and third term are of the order $c^2M^2\tau^3$ and this is by a factor $c\tau$ less than $\tilde{\tilde{\Gamma}}_1$. Hence

$$2\pi I(\omega) = \frac{c}{\omega^2}\,\tilde{\tilde{\Gamma}}_1 \quad \text{for} \quad |\omega| \gtrapprox \frac{1}{\tau}\,.$$

The expression $\frac{c}{\omega^2}\tilde{\tilde{\Gamma}}_1$ seems to be the one-particle-approximation of the quasistatic approximation. Assume e.g.

$$h(\varphi(\zeta)) = \begin{cases} \text{const} = \omega_p(\zeta) & \text{for} \quad |t| \leqslant \frac{\tau}{2} \\ 0 & \text{for} \quad |t| > \frac{\tau}{2} \end{cases}$$

Then

$$\frac{c}{\omega^2}\bar{\bar{\Gamma}}_1 = Ec\,\frac{\omega_p(\zeta)^2}{\omega^2}\;\frac{4\sin^2(\omega-\omega_p(\zeta))\tau/2}{(\omega-\omega_p)^2}$$

Approximately

$$\frac{4\sin^2(\omega-\omega_p(\zeta))\tau/2}{(\omega-\omega_p)^2} \approx 2\pi\tau\delta(\omega-\omega_p(\zeta))$$

and finally

$$\begin{aligned} I(\omega)d\omega &\approx c\tau E\,\delta(\omega-\omega_p(\zeta))d\omega \\ &= c\tau\,\mathrm{Prob}\,\{\omega_p \in \omega,\ \omega+d\omega\}. \end{aligned}$$

The perturbation frequency is equal to zero, if the distance $|t - t_\iota|$ to the next impact time t_ι is $> \frac{\tau}{2}$ and is equal to $\omega_p(\zeta)$ in the other case, which has the probability $1 - e^{-c\tau} \approx c\tau$. So

(10) $$I(\omega)d\omega \approx \mathrm{Prob}\,\{X \in \omega,\ \omega + d\omega\} \quad \text{for} \quad \omega \gtrsim \frac{1}{\tau}.$$

II. Mathematical treatment of the problem

II.1. The first formula

Recall the general situation met in I. There was given a finite dimensional Hilbert space $\mathfrak{h} = \mathfrak{h}_a \otimes \mathfrak{h}_b$, a real vector space $\mathcal{M} = \mathbb{R}^3$ and a continuous mapping h from $\mathcal{M}$ into the Hermitian elements of $\mathcal{L}(\mathfrak{h})$, the space of linear applications of $\mathfrak{h}$ into itself. The function h had the property $h(0) = 0$. Then we had a function $\varphi(\zeta,t)$ with values in $\mathcal{M}$. Let us assume that ζ is taken from a compact space Z and that

$$\varphi : Z \times \mathbb{R} \to \mathcal{M}$$

is continuous and has the property

$$\varphi(\zeta,t) = 0 \quad \text{for} \quad |t| > \frac{\tau}{2} .$$

On Z there was given a probability distribution p_Z and on $\mathbb{R}_+$ the exponential distribution

$$p_c : p_c\{\xi,\ \xi + d\xi\} = cd\xi e^{-c\xi} \qquad \text{for} \quad c > 0.$$

Then we had two sequences of independent random variables $\zeta_1, \zeta_2, \ldots$ and $u_1, u_2, \ldots$ taking values in Z resp. $\mathbb{R}_+$ and being distributed with respect to the laws p_Z and p_c. Then $I(\omega)$ was given by

$$I(\omega) = \lim_{N\to\infty} \frac{c}{2\pi N} E\left(\int e^{-i\omega t}\, W_N(t)dt\right)\left(\int e^{-i\omega t}\, W_N(t)dt\right)^*$$

with

$$W_N(t) = \overset{t}{\underset{-\infty}{\tilde{\omega}}} \Phi_N - \mathbb{1}_{\leqslant 0} - \mathbb{1}_{> u_1+\ldots+u_N} \overset{+\infty}{\underset{-\infty}{\tilde{\omega}}} \Phi_N$$

(1)

$$\Phi_N = h(\varphi(\zeta_1) + \varphi(\zeta_2)\,[u_1] + \ldots + \varphi(\zeta_N)\,[u_1 + \ldots + u_{N-1}]).$$

By mathematical reasons we prefer the discussion of $R(t) = \int I(\omega)\, e^{i\omega t}\, d\omega$ instead of that of $I(\omega)$.

We have

$$R(t) = \lim_{N\to\infty} \frac{c}{N} E\, W_N * W_N^+.$$

We generalize and stop to assume that ζ_1 and u_1, ζ_2 and $u_2, \ldots$ are independent. Instead of this we assume that the pairs $(\zeta_1, u_1), (\zeta_2, u_2), \ldots$ are independent and identically distributed with respect to a law p on $Z \times \mathbb{R}_+$. We assume that

$$\text{(i) } 0 < Eu = \int p(d\zeta, du)\, u < \infty$$

$$\text{(ii) } p\{u > \tau\} > 0.$$

So our problem gets to investigate

$$R_N = \frac{1}{NEu} E\, W_N * W_N^+ \tag{2}$$

for $N \longrightarrow \infty$.

We shall at first establish the convergence of R_N and a formula for its limit, which is valid for any value of τ, but depends on τ. Then we are going to prove another formula which is independent of τ, but can only be verified for small τ.

Throughout the following considerations we shall need some functions which depend on intervals. Let $\alpha = (k, \ldots, \ell)$, $k \leq \ell$ be an interval of $\mathbb{N} = \{1, 2, \ldots\}$, the set of natural numbers, then we define

$$U(\alpha) = u_k + \ldots + u_\ell$$

$$\Phi(\alpha) = h(\varphi(\zeta_k) + \varphi(\zeta_{k+1})[u_k] + \ldots + \varphi(\zeta_\ell)[u_k + \ldots + u_{\ell-1}])$$

$$\alpha(\alpha) = \delta_{U(\alpha)} \overset{+\infty}{\underset{-\infty}{\tilde{\omega}}} \Phi(\alpha)$$

$$\beta(\alpha) = \overset{t}{\underset{-\infty}{\tilde{\omega}}} \Phi(\alpha) - 1_{<0} - 1_{>U(\alpha)} \overset{+\infty}{\underset{-\infty}{\tilde{\omega}}} \Phi(\alpha)$$

$$\bar{\beta}(\alpha) = \alpha(\alpha) * \beta(\alpha)^+$$

$$\bar{\bar{\beta}}(\alpha) = \beta(\alpha) * \beta(\alpha)^+ = \bar{\beta}(\alpha)^+ * \bar{\beta}(\alpha)$$

because $\alpha(\alpha)*\alpha(\alpha)^+ = \delta$, where δ is the Dirac measure at point 0, and $\delta_{U(\alpha)}$ is the Dirac measure at point $U(\alpha)$. Convolution with $\delta_{U(\alpha)}$ is equivalent to translation by $U(\alpha)$. We can write

$$R_N = \frac{1}{N} E\, \bar{\bar{\beta}}(1, 2, \ldots, N).$$

The following lemma will heavily be used throughout the paper.

Lemma 1. Let $k \leqslant m < \ell$ and $u_m > \tau$, then for $\sigma = (k, \ldots, \ell)$, $\sigma_1 = (k, \ldots, m)$, $\sigma_2 = (m+1, \ldots, \ell)$ we have

$$
\begin{aligned}
U(\sigma) &= U(\sigma_1) + U(\sigma_2) \\
\Phi(\sigma) &= \Phi(\sigma_1) + \Phi(\sigma_2)[U(\sigma_1)] \\
\alpha(\sigma) &= \alpha(\sigma_2) * \alpha(\sigma_1) \\
\beta(\sigma) &= \beta(\sigma_2) * \alpha(\sigma_1) + \beta(\sigma_1) \\
\tilde\beta(\sigma) &= \tilde\beta(\sigma_2) + \alpha(\sigma_2) * \tilde\beta(\sigma_1) \\
\tilde{\tilde\beta}(\sigma) &= \tilde{\tilde\beta}(\sigma_1) + \tilde{\tilde\beta}(\sigma_2) + \beta(\sigma_2) * \tilde\beta(\sigma_1) + \tilde\beta(\sigma_1)^+ * \beta(\sigma_2)^+ .
\end{aligned}
$$

Proof: Let t_o be such that

$$u_k + \ldots + u_{m-1} + \tau/2 < t_o < u_k + \ldots + u_m - \tau/2 .$$

Then the support of $\varphi(\zeta_k) + \ldots + (\zeta_m)[u_k + \ldots u_{m-1}]$ is contained in $]-\infty, t_o[$ and the support of $\varphi(\zeta_{m+1})[u_k + \ldots + u_{m+1}] + \ldots + \varphi(\zeta_k)[u_k + \ldots + u_{\ell-1}]$ is contained in $]t_o, \infty[$.

As $h(0) = 0$

$$
\begin{aligned}
\varphi(\sigma) &= h(\varphi(\zeta_k) + \ldots + \varphi(\zeta_m)[u_k + \ldots + u_{m-1}]) \\
&\quad + h(\varphi(\zeta_{m+1})[u_k + \ldots + u_m] + \ldots + \varphi(\zeta_\ell)[u_k + \ldots + u_\ell]) \\
&= \Phi(\sigma_1) + \Phi(\sigma_2)[U(\sigma_1)].
\end{aligned}
$$

In the same way

$$\tilde{\omega}_{-\infty}^{+\infty} \Phi(\sigma) = \bar{\omega}_{-\infty}^{+\infty} \Phi(\sigma_2) \, \bar{\omega}_{-\infty}^{+\infty} \Phi(\sigma_1)$$

hence

$$\alpha(\sigma) = \alpha(\sigma_2) * \alpha(\sigma_1).$$

In order to prove the equation for $\beta(\mathscr{A})$, one observes

$$\overset{t}{\underset{-\infty}{\tilde{\omega}}} \Phi(\mathscr{A}) = \begin{cases} \overset{t}{\underset{-\infty}{\tilde{\omega}}} \Phi(\mathscr{A}_2)\left[U(\mathscr{A}_1)\right] \overset{+\infty}{\underset{-\infty}{\tilde{\omega}}} \Phi(\mathscr{A}_1) & \text{for } t \geqslant t_o \\ \overset{t}{\underset{-\infty}{\tilde{\omega}}} \Phi(\mathscr{A}_1) & \text{for } t \leqslant t_o \end{cases}$$

From the equation for $\beta(\mathscr{A})$, and $\alpha(\mathscr{A})$, one deduces the equations for $\tilde{\beta}(\mathscr{A})$ and $\tilde{\tilde{\beta}}(\mathscr{A})$.

$$\begin{aligned} \tilde{\beta} &= \alpha\mathscr{A} * \beta\mathscr{A}^+ \\ &= \alpha\mathscr{A}_2 * \alpha\mathscr{A}_1 * ((\alpha\mathscr{A}_1)^+ * \beta\mathscr{A}_2^+ + \beta\mathscr{A}_1^+) \\ &= \alpha\mathscr{A}_2 * \beta\mathscr{A}_2^+ + \alpha\mathscr{A}_2 * \alpha\mathscr{A}_1 * \beta\mathscr{A}_1^+ \\ &= \tilde{\beta}\mathscr{A}_2 + \alpha\mathscr{A}_2 * \tilde{\beta}\mathscr{A}_1 . \end{aligned}$$

$$\begin{aligned} \tilde{\tilde{\beta}}\mathscr{A} &= \beta\mathscr{A} * \beta\mathscr{A}^+ \\ &= (\beta\mathscr{A}_2 * \alpha\mathscr{A}_1 + \beta\mathscr{A}_1) * (\beta\mathscr{A}_1^+ + \alpha\mathscr{A}_1^+ * \beta\mathscr{A}_2^+) \\ &= \tilde{\tilde{\beta}}\mathscr{A}_1 + \beta\mathscr{A}_2 + \beta\mathscr{A}_2 * \tilde{\beta}\mathscr{A}_1 + \tilde{\beta}\mathscr{A}_1^+ * \beta\mathscr{A}_2^+ . \end{aligned}$$

This proves the lemma.

<u>Corollary.</u> Let $k_o \leqslant k_1 < k_2 < \ldots < k_{p-1} < k_p$ and $u_{k_1} > \tau$, $u_{k_2} > \tau, \ldots, u_{k_{p-1}} > \tau$. Then for

$$\begin{aligned} \mathscr{A} &= (k_o, \ldots, k_p) \\ \mathscr{A}_1 &= (k_o, \ldots, k_1) \\ \mathscr{A}_2 &= (k_1+1, \ldots, k_2) \\ &\vdots \\ \mathscr{A}_p &= (k_{p-1}+1, \ldots, k_p) \end{aligned}$$

we have

$$U(\mathscr{A}) = U(\mathscr{A}_1) + \ldots + U(\mathscr{A}_p)$$

$$\Phi(\mathscr{A}) = \Phi(\mathscr{A}_1) + \Phi(\mathscr{A}_2)\left[U(\mathscr{A}_1)\right] + \ldots + \Phi(\mathscr{A}_p)\left[U(\mathscr{A}_1) + \ldots + U(\mathscr{A}_{p-1})\right]$$

$$\alpha(\mathscr{A}) = \alpha(\mathscr{A}_p) * \alpha(\mathscr{A}_{p-1}) * \ldots * \alpha(\mathscr{A}_1)$$

$$\beta(\alpha) = \beta(\alpha_p) * \alpha(\alpha_{p-1}) * \ldots * \alpha(\alpha_1)$$
$$+ \beta(\alpha_{p-1}) * \alpha(\alpha_{p-2}) * \ldots * \alpha(\alpha_1)$$
$$+ \ldots + \beta(\alpha_1)$$

$$\tilde{\beta}(\alpha) = \alpha(\alpha_p) * \alpha(\alpha_{p-1}) * \ldots * \alpha(\alpha_2) * \tilde{\beta}(\alpha_1)$$
$$+ \alpha(\alpha_p) * \alpha(\alpha_{p-1}) * \ldots * \tilde{\beta}(\alpha_2)$$
$$+ \ldots + \tilde{\beta}(\alpha_p).$$

$$\tilde{\tilde{\beta}}(\ ) = \sum_{j=1}^{p} \tilde{\tilde{\beta}}(\alpha_j) + \sum_{1 \leq i < j \leq p} \beta(\alpha_j) * \alpha(\alpha_{j-1}) * \ldots$$
$$\ldots * \alpha(\alpha_{i+1}) * \tilde{\beta}(\alpha_i) + \sum_{1 < i < j \leq p} \tilde{\beta}(\alpha_i)^+ * \alpha(\alpha_{i+1})^+ * \ldots$$
$$\ldots \alpha(\alpha_{j-1})^+ * \beta(\alpha_j)^+.$$

The first method can be described as the method of the stochastic interval. We introduce the compact space $\mathfrak{z} = (Z \times \overline{\mathbb{R}}_+)^{\mathbb{N}}$, $\overline{\mathbb{R}}_+ = \mathbb{R}_+ \cup \{\infty\}$, $\mathbb{N} = \{1, 2, \ldots\}$ and on $\mathfrak{z}$ the product measure $P = p^{\otimes N}$ where p is the measure p extended from $Z \times \mathbb{R}_+$ to $Z \times \overline{\mathbb{R}}_+$. The space $\mathfrak{z}$ is the set of all infinite sequences $((\zeta'_j, u'_j))_{j \in \mathbb{N}}$ with $(\zeta'_j, u_j) \in Z \times \overline{\mathbb{R}}_+$.

Define by $\sum$ the space of all finite sequences $((\zeta'_j, u'_j))_{j \in \mathbb{N}}$, i.e. $\sum$ is the topological sum

$$\sum = \bigcup_{k=1}^{\infty} (Z \times \overline{\mathbb{R}}_+)^k$$

The space $\sum$ is locally compact.

The function $K : \sigma' = ((\zeta_1', u_1'), \ldots, (\zeta_k', u_k')) \mapsto k$ is continuous on $\sum$. Any continuous function f on $\sum$ can be decomposed into

$$f = \sum_{k=1}^{\infty} f\, \mathbb{1}\{K = k\} = \sum_{k=1}^{\infty} f_k$$

where $\mathbb{1}$ denotes again the characteristic function.

We define applications $\sigma_1, \sigma_2, \ldots$ from $\mathfrak{z}$ into $\sum$. Let $\mathfrak{z}' = ((\zeta_1', u_1'), (\zeta_2', u_2'), \ldots) \in \mathfrak{z}$ and let k_1 be the first integer j with $u_j > \tau$, let k_2 be the second integer with $u_j > \tau$,.... Then

$$\sigma_1(\mathfrak{z}') = ((\zeta_1', u_1'), \ldots, (\zeta_{k_1}', u_{k_1}'))$$

$$\sigma_2(\mathfrak{z}') = ((\zeta_{k_1+1}', u_{k_1+1}'), \ldots, (\zeta_{k_2}', u_{k_2}'))$$

$$\vdots$$

Lemma 2. The applications $\sigma_1, \sigma_2, \ldots$ from $\mathfrak{z}$ to $\sum$ are P-measurable and independent. The image of P with respect to each of them is the measure Q on $\sum$ given by

$$\langle Q , f \rangle = \sum_{k=1}^{\infty} \int_{u_1' \leqslant \tau} p(\zeta_1', u_1') \ldots \int_{u_{k-1}' \leqslant \tau} p(\zeta_{k_1-1}' u_{k-1}') \int_{u_k' > \tau} p(\zeta_{k_1}' u_k')$$

$$f_k((\zeta_1', u_1'), \ldots, (\zeta_k', u_k'))$$

if

$$f = \sum_{k=1}^{\infty} f_k = \sum_{k=1}^{\infty} f \cdot \mathbb{1}\{K = k\}$$

is a continuous function of compact support in $\sum$.

Proof. Fix a natural number ℓ and define

$$S_\ell : \mathfrak{z} \mapsto \textstyle\sum^\ell$$

$$\mathfrak{z}' \mapsto (\sigma_1, \sigma_2, \ldots, \sigma_\ell).$$

Call $K_j = K \circ \sigma_j$, $j = 1, \ldots, \ell$, i.e. K_j is the length of σ_j. Then $K_1 + \ldots + K_\ell \geqslant m$ is equivalent to the statement that among the first m pairs (ζ_j', u_j') of $\mathfrak{z}'$ there are ℓ pairs with $u_j > \tau$. Therefore given any $\varepsilon > 0$ we can find an m with $P\{K_1 + \ldots + K_\ell \geqslant m\} < \varepsilon$. Then choose $\delta > 0$ such that the set of all $\mathfrak{z}' \in \mathfrak{Z}$ with $u_j' \notin]\tau, \tau + \delta[$ for $j = 1, \ldots, m$ has P-measure $\geqslant 1 - \varepsilon$. The set $\mathfrak{K}$ of all $\mathfrak{z}' \in \mathfrak{Z}$ with the properties (1) $u_j \notin]\tau, \tau + \delta[$ for $j = 1, \ldots, m$ (2) there are $\geqslant \ell$ pairs (ζ_j', u_j'), $j = 1, \ldots, m$ with $u_j' \geqslant \tau + \delta$ is compact in $\mathfrak{Z}$ and has P-measure $\geqslant 1 - 2\varepsilon$. The function S_ℓ is continuous on $\check{\mathfrak{K}}$. Hence S_ℓ is P-measurable.

Let $f^{(1)}, f^{(2)}, \ldots, f^{(\ell)}$ be continuous functions with compact support on $\sum$. Then

$$Ef^{(1)}(\sigma_1)f^{(2)}(\sigma_2) \ldots f^{(\ell)}(\sigma_\ell)$$

$$= E \sum_{k_1,\ldots,k_\ell=1}^{\infty} \prod_{j=1}^{\ell} \Big[f_{k_j}^{(j)} ((\zeta_{k_{j-1}+1}, u_{k_{j-1}+1}), \ldots, (\zeta_{k_j}, u_{k_j})) \cdot$$

$$\cdot \mathbb{1}\{u_{k_{j-1}+1} \leqslant \tau\} \ldots \mathbb{1}\{u_{k_{j-1}} \leqslant \tau\} \mathbb{1}\{u_{k_j} > \tau\} \Big]$$

$$= \sum_{k_1,\ldots,k_\ell=1}^{\infty} \prod_{j=1}^{\ell} E \Big[\quad \Big]$$

$$= \sum_{k_1,\ldots,k_\ell=1}^{\infty} \prod_{j=1}^{\ell} E \Big[f_{k_j}^{(j)} ((\zeta_1, u_1), \ldots, (\zeta_{k_j}, u_{k_j})) \cdot$$

$$\cdot \mathbb{1}\{u_1 \leqslant \tau\} \ldots \mathbb{1}\{u_{k_{j-1}} \leqslant \tau\} \mathbb{1}\{u_{k_j} > \tau\} \Big]$$

$$= \prod_{j=1}^{\ell} \sum_{k=1}^{\infty} Ef_k^{(j)} ((\zeta_1, u_1), \ldots, (\zeta_k, u_k)) \mathbb{1}\{u_1 \leqslant \tau\} \ldots \mathbb{1}\{u_{k-1} \leqslant \tau\} \cdot$$

$$\cdot \mathbb{1}\{u_k > \tau\}$$

$$= \prod_{j=1}^{\ell} \langle Q, f^{(j)} \rangle \qquad \text{q.e.d.}$$

During the proof of the last lemma we called $K_j = K \circ \sigma_j$. The lemma shows that the K_j's are independent and distributed with respect to the same law.

$$P\{K_j = k\} = p\{u \leq \tau\}^{k-1} \cdot p\{u > \tau\}.$$

The expectation of K_j is

$$EK_j = \frac{1}{p\{u > \tau\}} < \infty$$

by assumption (ii), $p\{u > \tau\} > 0$ made above.
The function $U = u_1 + \ldots + u_{K_1}$ has the expectation

$$EU = \frac{Eu}{p\{u > \tau\}}.$$

Call $N_l = K_1 + \ldots + K_{\ell-1}$. Then $EN_\ell = (\ell-1)/p\{u > \tau\}$. This makes the following lemma plausible.

Lemma 3. The functions

$$R_N = \frac{1}{Eu \cdot N} E\, W_N * W_N^+, \; N \longrightarrow \infty$$

converge uniformly in every compact interval, iff

$$P_\ell = \frac{1}{EU(\ell-1)} E\, W_{N_\ell} * W_{N_\ell}^+$$

converge in every compact interval. If both converge, the limits are equal.

Proof. Put $a = p\{u > \tau\}$. If $a = 1$, then $K_j = 1$, for $j = 1, 2, \ldots$ and both formulae are identical. We assume that $a < 1$.

Let $N' \geq N$ be two integers. Then $\Phi_N(t)$ and $\Phi_{N'}(t)$ coincide for $t \leq u_1 + \ldots + u_{N'-1} - \tau/2$, hence W_N and $W_{N'}$ are equal outside the interval $[u_1 + \ldots + u_{N'-1} - \tau/2, u_1 + \ldots + u_N + \tau/2]$. As $|W_N|$ and $|W_{N'}|$ are bounded by 2 we have at first

$$\int |W_N(t) - W_{N'}(t)| dt \leq 4(u_{N'} + u_{N'+1} + \ldots + u_N + \tau)$$

and then

(1) $|W_N * W_N^+ - W_{N'} * W_{N'}^+| \leqslant 32(u_{N'} + \dots + u_N + \tau)$.

Let $[\ell/a]$ the smallest integer $\geqslant \ell/a$. Then

$$|E(W_{[\ell/a]} * W_{[\ell/a]}^+ - W_{N_\ell} * W_{N_\ell}^+)|$$

$$\leqslant 32 \sum_{n=1}^{[\ell/a]} E(u_n + \dots + u_{[\ell/a]} + \tau)\, 1\{N_\ell = n\}$$

$$+ 32 \sum_{[\ell/a]+1}^{\infty} E(u_{[\ell/a]+1} + \dots + u_n + \tau)\, 1\{N_\ell = n\}.$$

Now we have

$$Eu_j\, 1\{N_\ell = n\} = Eu_j E\, 1\{N_\ell = n\} = Eu \cdot P\{N_\ell = n\}$$

for $j > n$ and for $j \leqslant n$

$$Eu_j\, 1\{N_\ell = n\} = \sum_{k_1 + \dots + k_{\ell-1} = n} Eu_j\, 1\{K_1 = k_1\} \dots 1\{K_{\ell-1} = k_{\ell-1}\}$$

and

$$Eu_j\, 1\{K_1 = k_1\} \dots 1\{K_{\ell-1} = k_{\ell-1}\}$$

$$= \begin{cases} a^{\ell-2}(1-a)^{n-\ell+1}\, Eu\, 1\{u > \tau\} & \text{if } j \in \{k_1, \dots, k_{\ell-1}\} \\ a^{\ell-1}(1-a)^{n-\ell}\, Eu\, 1\{u \leqslant \tau\} & \text{if } j \notin \{k_1, \dots, k_{\ell-1}\} \end{cases}$$

$$\leqslant \frac{1}{\min(a, 1-a)}\, Eu \cdot a^{\ell-1}(1-a)^{n-\ell+1}.$$

Hence

$$Eu_j\, 1\{N_\ell = n\} \leqslant \frac{1}{\min(a, 1-a)}\, Eu \cdot P\{N_\ell = n\}.$$

and finally

$$\frac{1}{(\ell-1)EU} E \left| W_{[\ell/a]} * W^{+}_{[\ell/a]} - W_{N_\ell} * W^{+}_{N_\ell} \right|$$

$$\leqslant \frac{32a}{\min(a,\ 1-a)} \frac{1}{\ell-1} (1 + \tau + E|N_\ell - [\ell/a]|)$$

$$\longrightarrow 0 \quad \text{uniformly in } \mathbb{R}.$$

By easy reasoning there follows from (1) that convergence of

$\frac{1}{(\ell-1)EU} EW_{[\ell/a]} * W^{+}_{[\ell/a]}$ implies convergence of $1/_N\, EW_N * W^{+}_N$.

This completes the proof of the lemma.

The sequences $\sigma_1, \sigma_2, \ldots$ are characterized in a unique way by the intervals $(1, \ldots, K_1), (K_1 + 1, \ldots, K_2), \ldots$, which we may call $\mathfrak{a}(\sigma_1), \mathfrak{a}(\sigma_2), \ldots$. Instead of $\Phi(\mathfrak{a}(\sigma_j))$ we write for short $\Phi(\sigma_j)$ and use this abbreviation for $U, \alpha, \beta, \tilde{\beta}, \tilde{\tilde{\beta}}$, too. Denote by $\sigma_1 \circ \sigma_2 \circ \ldots \circ \sigma_{\ell-1}$ the combined sequence which is characterized by the interval $(1, \ldots, K_1 + \ldots + K_{\ell-1}) = (1, \ldots, N)$. Then

$$P_\ell = \frac{1}{(\ell-1)EU} E\tilde{\tilde{\beta}}(\sigma_1 \circ \ldots \circ \sigma_\ell).$$

<u>Lemma 4.</u> For $\ell \longrightarrow \infty$ the functions P_ℓ converge uniformly on every compact interval of $\mathbb{R}$ to

$$P = \frac{1}{EU} (\tilde{\tilde{\vartheta}} + \vartheta * (\delta - \eta)^{*-1} * \tilde{\vartheta}$$
$$+ \tilde{\vartheta}^{+} * (\delta - \eta^{+})^{*-1} * \delta^{+})$$

with

$$\eta = E\alpha(\sigma), \quad \tilde{\vartheta} = E\tilde{\beta}(\sigma)$$
$$\vartheta = E\beta(\sigma), \quad \tilde{\tilde{\vartheta}} = E\tilde{\tilde{\beta}}(\sigma)$$

and σ is a stochastic variable on $\sum$ distributed with respect to Q.

<u>Proof.</u> By the corollary of lemma 1 one obtains

$$\tilde{\tilde{\beta}}(\sigma_1 \circ \cdots \circ \sigma_\ell) = \sum_j \tilde{\tilde{\beta}}(\sigma_j) +$$

$$+ \sum_{1 \leq i < j \leq \ell} \Big[\beta(\sigma_j) * \alpha(\sigma_{j-1}) * \ldots * \alpha(\sigma_{i+1}) * \tilde{\beta}(\sigma_i) +$$

$$+ \tilde{\beta}(\sigma_i)^+ * \alpha(\sigma_{i+1})^+ * \ldots * \alpha(\sigma_{j-1})^+ * \beta(\sigma_j)^+ \Big]$$

and as the σ_j's are independent and identically distributed

$$E\tilde{\tilde{\beta}}(\sigma_1 \circ \ldots \circ \sigma_\ell) = \ell\, E\tilde{\tilde{\beta}}(\sigma) +$$

$$+ \Big[\sum_{1 \leq i < j \leq \ell} E\beta(\sigma) * (E\alpha(\sigma))^{*(j-i-1)} * E\tilde{\beta}(\sigma) \Big] + \Big[\quad \Big]^+$$

$$= \ell \tilde{\tilde{\vartheta}} + \Big[\sum_{k=1}^{\ell-1} (\vartheta * \eta^{*(k-1)} * \tilde{\vartheta}) \cdot (\ell - k) \Big] + \Big[\quad \Big]^+$$

and by formal reasoning

$$EU \cdot P_{\ell+1} = \frac{1}{\ell} E\tilde{\tilde{\beta}}(\sigma_1 \circ \ldots \circ \sigma_\ell)$$

$$\longrightarrow \tilde{\tilde{\vartheta}} + \Big[\sum_{k=1}^{\infty} \vartheta * \eta^{*(k-1)} * \tilde{\vartheta} \Big] + \Big[\quad \Big]^+$$

$$= \tilde{\tilde{\vartheta}} + \big[\vartheta * (\delta - \eta)^{*-1} * \tilde{\vartheta} \big] + [\ ]^+.$$

There are several gaps in this reasoning. We shall fill them up.

How to define

$$\eta = E \underset{-\infty}{\overset{+\infty}{\tilde{\omega}}} \Phi(\sigma)\delta_{U(\sigma)} \quad ?$$

Let $\varphi \in C_*(\mathbb{R})$ be the space of all continuous real valued functions on $\mathbb{R}$ with compact support. Then

$$\varphi \longmapsto E \Big\langle \underset{-\infty}{\overset{+\infty}{\tilde{\omega}}} \Phi(\sigma)\delta_{U(\sigma)}, \varphi \Big\rangle$$

$$= E \mathop{\varpi}_{-\infty}^{+\infty} \Phi(\sigma)\varphi(U(\sigma))$$

defines a $\mathcal{L}(\mathcal{H})$-valued measure on $\mathbb{R}$ with support in $\mathbb{R}_+$ which we denote by

$$E \mathop{\varpi}_{-\infty}^{+\infty} \Phi(\sigma)\delta_{U(\sigma)}.$$

The following lemma shows that commuting of expectations and convolution is legitimate in our case.

Lemma 5. Let X_1, X_2 be two locally compact Hausdorff spaces, let for $i = 1, 2$ be p_i a probability measure on X_i, let U_i be a p_i-measurable application from X_i into $\overline{\mathbb{R}}_+$ such that $p_i\{U_i = \infty\} = 0$, and A_i be a p_i-integrable function from X_i into $\mathcal{L}(\mathcal{H})$. Let F_1 be an integrable continuous application from X_1 into the Banach space $C_0(\mathbb{R})$ of all continuous functions on $\mathbb{R}$ vanishing at infinity. Then

$$\iint p_1 p_2 A_1 A_2 \delta_{U_1 U_2} = \left(\int p_1 A_1 \delta_{U_1}\right) * \left(\int p_2 A_2 \delta_{U_2}\right)$$

and

$$\iint p_1 p_2 F_1 * A_2 \delta_{U_2} = \int p_1 F_1 * \int p_2 A_2 \delta_{U_2} .$$

Proof. Let $\Psi \in C_*(\mathbb{R} \times \mathbb{R})$ and call

$$\int p_i A_i \delta_{U_i} = M_i.$$

Then

$$\left\langle \iint p_1 p_2 A_1 A_2 \delta_{U_1}(\xi)\delta_{U_2}(\eta), \Psi(\xi, \eta)\right\rangle$$
$$= \iint p_1 p_2 A_1 A_2 \Psi(U_1, U_2)$$
$$= \iint M_1(\xi) M_2(\eta) \Psi(\xi, \eta).$$

This equation holdsobviously, if $\Psi(\xi,\eta) = \Psi_1(\xi)\Psi_2(\eta)$. A density argument yields the formula for all Ψ.

Take now $\Psi(\xi,\eta) = \chi(\xi + \eta)$, $\chi \in C_*(\mathbb{R}_+)$ then by easy extension of the last equation

$$\Big\langle \iint p_1 p_2 A_1 A_2 \delta_{U_1 + U_2}(\xi),\ \chi(\xi) \Big\rangle$$
$$= \iint p_1 p_2 A_1 A_2\, \chi(U_1 + U_2)$$
$$= \iint M_1(\xi) M_2(\eta)\, \chi(\xi + \eta)$$
$$= \langle M_1 * M_2,\ \chi \rangle .$$

This proves the first equation of the lemma.

We have

$$\iint p_1 p_2 (F_1 * A_2 \delta_{U_2})$$
$$= \iint p_1(x_1) p_2(x_2) F_1(x_1, \xi - U(x_2)) A_2(x_2).$$

As $(x_1, x_2) \mapsto (F_1(x_1, \xi - U(x_2))$ is $p_1 \otimes p_2$-measurable, we apply Fubini's theorem and get

$$= \int p_2(x_2) A_2(x_2) \int p_1(x_1) F_1(x_1, \xi - U(x_2))$$
$$= \int p_2(x_2) A_2(x_2) \Big(\int p_1 F_1\Big)(\xi - U(x_2))$$
$$= M * \Big(\int p_1 F_1\Big)(\xi).$$

We want to apply lemma 5 e.g. in the following situation

$$E\beta(\sigma_2) * \alpha(\sigma_1) = E\beta(\sigma_2) * E\alpha(\sigma_1)$$

with

$$\beta(\sigma) = \overset{t}{\underset{-\infty}{\tilde{\omega}}} \Phi(\sigma) - \mathbb{1}_{\leqslant 0} - \mathbb{1}_{> U(0)} \overset{+\infty}{\underset{-\infty}{\tilde{\omega}}} \Phi(\sigma)$$
$$= \overset{t}{\underset{-\infty}{\tilde{\omega}}} \Phi(\sigma) - \mathbb{1}_{\leqslant 0} * \gamma - (\mathbb{1}_{> U(\sigma)}) \overset{+\infty}{\underset{-\infty}{\tilde{\omega}}} \Phi(\sigma)$$
$$+ (\mathbb{1}_{\leqslant 0} * \gamma - \mathbb{1}_{\leqslant 0}) + (\mathbb{1}_{> 0} * \gamma - \mathbb{1}_{> 0}) * \delta_{U(\sigma)} \overset{+\infty}{\underset{-\infty}{\tilde{\omega}}} \Phi(\sigma)$$

where $\gamma \geqslant 0$ is a continuous function of small support and $\int \gamma = 1$. The function $\beta(\sigma)$ decomposes into three terms, the first one is a continuous function $\in C_0(\mathbb{R}, \mathcal{L}(\mathcal{H}))$ depending in a continuous way on σ. So we can apply the second formula of lemma 5. The second term is a piecewise continuous function which does not depend on σ at all. For the third one we have to apply the first formula of lemma 5.

Another gap in the proof of lemma 4 is the way how the sum

$$\sum_{k=1}^{\infty} n^{*(k-1)}$$

converges. This will be clarified by lemma 6. Let μ be an $\mathcal{L}(\mathfrak{H})$-valued measure on $\mathbb{R}_+$ and let $\xi \geq 0$.

We define

$$\|\mu\|_\xi = \sup\{|\langle \mu,\phi\rangle| : \phi \in C_*(\mathbb{R}_+), \; |\phi| \leq 1, \sup \phi \subset [0,\xi]\}$$

where $|\langle \mu,\phi\rangle|$ is the $\mathcal{L}(\mathfrak{H})$-norm of $\langle \mu,\phi\rangle$.

Lemma 6. Let μ be an $\mathcal{L}(\mathfrak{H})$-valued measure on $\mathbb{R}_+$ with $|\mu|\{0\}<1$. Then the series

$$\delta + \mu + \mu^{*2} + \dots$$

converges in $\|\cdot\|_\xi$-norm for every $\xi > 0$ and one has

$$(\delta - \mu) * (\delta + \mu + \mu^{*2} + \dots) = \delta$$

or

$$\delta + \mu + \mu^{*2} + \dots = (\delta - \mu)^{*-1} .$$

Proof. Choose $\varepsilon > 0$ such that $|\mu|[0,\varepsilon] = \varkappa < 1$. Then $\mu = \mu_1 + \mu_2 = \mu \mathbb{1}[0,\varepsilon] + \mu \mathbb{1}]\varepsilon,\infty[$. The measure μ_2^{*N} has its support in $[N\varepsilon,\infty[$, so $\|\mu_2^{*N}\|_\xi = 0$ for $N\varepsilon \geq \xi$. Let N_0 be the smallest integer $\geq \xi/\varepsilon$. Then

$$\|\mu^{*N}\|_\xi \leq \sum_{k=0}^{N}\binom{N}{k}\|\mu_1\|_\xi^{N-k} \; \|\mu_2\|_\xi^k$$

$$\leq \sum_{k=0}^{N_0}\binom{N}{k} \varkappa^{N-k} \|\mu_2\|_\xi^k$$

and

$$\sum_{N=1}^{\infty}\|\mu^{*N}\|_\xi \leq \sum_{k=0}^{N_0}\|\mu_2\|_\xi^k \sum_{N=0}^{\infty}\binom{N}{k}\varkappa^{N-k}$$

$$= \sum_{k=0}^{N_0}\|\mu_2\|_\xi^k \frac{1}{(1-\varkappa)^k} < \infty.$$

This inequality implies the lemma.

In order to apply lemma 6 to the proof of lemma 4, we have to check that $|\eta|\{0\} < 1$. Now $\eta = E\delta_{U(\sigma)} \varpi_{-\infty}^{+\infty} \Phi(\sigma)$, therefore $|\eta|\{0\} \leqslant P\{U(\sigma) = 0\} < 1$, because $EU = Eu/p\{u > \tau\} > 0$. The kind of convergence of the sum $\sum_{k=1}^{\infty} \eta^{*(k-1)}$ implies the uniform convergence of P_ℓ on every compact interval of the line.

We combine all these results into

<u>Theorem 1.</u> For $N \longrightarrow 0$ there converges

$$R_N = \frac{1}{N\ Eu} E\ W_N * W_N^+$$

to

$$R = \frac{p\{u > \tau\}}{Eu} (\bar{\bar{\vartheta}} + \vartheta * (\delta - \eta)^{*-1} * \bar{\vartheta} + \bar{\vartheta}^+ * (\delta - \eta^+)^{*-1} * \vartheta^+)$$

where

$$\eta = E\sum_{j=1}^{\infty} \delta_{u_j} \varpi_{-\infty}^{+\infty} \Phi_j \cdot \mathbb{1}\{u_1 \leqslant \tau\} \ \ldots\ \mathbb{1}\{u_{j-1} \leqslant \tau\}\, \mathbb{1}\{u_j > \tau\}$$

$$\vartheta = E\sum_{j=1}^{\infty}\left(\varpi_{-\infty}^{t} \Phi_j - \mathbb{1}_{\leqslant 0}(t) - \mathbb{1}_{> U_j}(t) \varpi_{-\infty}^{+\infty} \Phi_j\right)$$

$$\mathbb{1}\{u_1 \leqslant \tau\} \ldots \mathbb{1}\{u_{j-1} \leqslant \tau\}\, \mathbb{1}\{u_j > \tau\}$$

$$\bar{\vartheta} = E\sum_{j=1}^{\infty}\left(\varpi_{u_j - t}^{\infty} \Phi_j - \mathbb{1}_{< 0}(t) - \mathbb{1}_{\geqslant U_j} \varpi_{-\infty}^{+\infty} \Phi_j\right)$$

$$\mathbb{1}\{u_1 \leqslant \tau\} \ldots \mathbb{1}\{u_{j-1} \leqslant \tau\}\, \mathbb{1}\{u_j > \tau\}$$

$$\bar{\bar{\vartheta}} = \sum_{j=1}^{\infty}\left(\varpi_{-\infty}^{t} \Phi_j - \mathbb{1}_{\leqslant 0} - \mathbb{1}_{> U_j} \varpi_{-\infty}^{+\infty} \Phi_j\right) *$$

$$\left(\varpi_{-\infty}^{t} \Phi_j - \mathbb{1}_{\leqslant 0} - \mathbb{1}_{> U_j} \varpi_{-\infty}^{+\infty} \Phi_j\right)^+$$

$$\mathbb{1}\{u_1 \leqslant \tau\} \ldots \mathbb{1}\{u_{j-1} \leqslant \tau\}\, \mathbb{1}\{u_j > \tau\}$$

and

$$\Phi_j = h(\varphi(\zeta_1) + \varphi(\zeta_2)[u_1] + \ldots + \varphi(\zeta_j)[u_1 + \ldots + u_{j-1}])$$

$$U_j = u_1 + \ldots + u_j .$$

This theorem gives R and hence $I(\omega)$ as functions of t- resp. ω-dependent quantities $\eta, \vartheta, \tilde{\vartheta}, \tilde{\tilde{\vartheta}}$ which are expressed by infinite sums. The first term of these sums is the one-particle-interaction, the second is the two-particle-interaction and so on. A nice fact of this formula is that its convergence has been shown in any case, but it has already been pointed out that the terms $\eta, \vartheta, \tilde{\vartheta}, \tilde{\tilde{\vartheta}}$ heavily depend on the choice of τ.

II.2. Algebraic considerations

Let $\mathbb{N}$ be the set of all natural numbers $\mathbb{N} = \{1, 2, \ldots\}$. We consider finite intervals $\mathfrak{a} = (k, k+1, \ldots, \ell)$ in $\mathbb{N}$. Call $\mathfrak{A}$ the set of all those intervals and $\mathfrak{F}$ the free real algebra generated by $\mathfrak{A}$. The algebra $\mathfrak{F}$ consists of all monomials

$$\mathfrak{m} = \mathfrak{a}_1 \mathfrak{a}_2 \ldots \mathfrak{a}_K, \quad \mathfrak{a}_i \in \mathfrak{A},$$

and the unit element 1 and all the real linear combinations of monomials.

We write

$$\mathfrak{a} = \mathfrak{a}_1 \circ \mathfrak{a}_2 \circ \ldots \circ \mathfrak{a}_K$$

if $\mathfrak{a}$ is the union of the <u>subsequent</u> intervals e.g.

$$\begin{aligned}
\mathfrak{a} &= (k, \ldots, \ell) \\
\mathfrak{a}_1 &= (k, \ldots, \ell_1) \\
\mathfrak{a}_2 &= (\ell_1 + 1, \ldots, \ell_2) \\
&\vdots \\
\mathfrak{a}_K &= (\ell_{K-1} + 1, \ldots, \ell_K = \ell)
\end{aligned}$$

Let $m \in \mathbb{N}$, $\alpha = (k, \ldots, \ell)$, then define

$$\gamma_m \alpha = \begin{cases} (k, \ldots, m)(m+1, \ldots, \ell), & \text{if } k \leqslant m < \ell \\ (k, \ldots, m) & \text{otherwise.} \end{cases}$$

γ_m is called a decomposition operator. If $k \leqslant m < \ell$, then we say that γ_m decomposes α. γ_m can be extended in a unique way to a homomorphism from $\mathcal{J}$ into $\mathcal{J}$ by setting

$$\gamma_m \alpha_1 \ldots \alpha_K = \gamma_m \alpha_1 \ldots \gamma_m \alpha_K.$$

This homomorphism γ_m is an idempotent operator on $\mathcal{J}$.

Let $\alpha = (k, \ldots, \ell)$, then we set

$$[\alpha] = [k, \ldots, \ell]$$

$$= \sum_{K \geqslant 1} (-1)^{K+1} \sum_{\alpha_1 \circ \ldots \circ \alpha_K = \alpha} \alpha_1 \alpha_2 \ldots \alpha_K .$$

The following lemma is formulated only for the special interval $(1, 2, \ldots, N)$, but it holds in a similar way for any other interval.

<u>Lemma 7.</u> $(1, 2, \ldots, N) = [1, 2, \ldots, N] +$

$+ (1) [2, \ldots, N] + (1,2) [3, \ldots, N] + \ldots + (1, 2, \ldots, N-1) [N]$

$= [1, 2, \ldots, N] + [1, \ldots, N-1] (1) + [1, \ldots, N-2] (N-1, N)$

$+ \ldots + [1] (2, \ldots, N).$

<u>Proof.</u> By definition

$$[1,2,\ldots N] = (1,2,\ldots,N) + \sum_{K \geqslant 2} \sum_{\alpha_1 \circ \ldots \circ \alpha_K = (1,2,\ldots,N)} (-1)^{K+1} \alpha_1 \ldots \alpha_K$$

$$= (1,2,\ldots,N) - \sum_{k=1}^{N-1} (1,\ldots,k) \sum_{K \geqslant 1} (-1)^{K+1} \sum_{\alpha_1 \circ \ldots \circ \alpha_K = (k+1,\ldots,N)} \alpha_1 \alpha_2 \ldots \alpha_K$$

$$= (1,2,\ldots,N) - \sum_{k=1}^{N-1} (1,\ldots,k) [k+1,\ldots,N].$$

This proves the first formula. The second one is derived in the same way.

Lemma 8. If $\mathscr{Y}_m$ decomposes $\mathfrak{a}$, then $\mathscr{Y}_m[\mathfrak{a}] = 0$.

Proof. We proceed by induction of the length of $\mathfrak{a}$. If the length $|\mathfrak{a}| = 1$, then all is clear. Assume the lemma proven for length $\leq N-1$ and check it for $\mathfrak{a} = (1, 2, \ldots, N)$ and $1 \leq m < N$. By the first formula of lemma 7

$$\begin{aligned}\mathscr{Y}_m(1,2,\ldots,N) &= (1,2,\ldots,m)(m+1,\ldots,N)\\ &= \mathscr{Y}_m[1,2,\ldots,N] + (1,2,\ldots,m)[m+1,\ldots,N]\\ &\quad + (1,2,\ldots,m)(m+1)[m+2,\ldots,N] + \ldots + (1,2,\ldots,m)(m+1,\ldots\\ &\qquad\qquad \ldots N-1)[N]\\ &= \mathscr{Y}_m[1,2,\ldots,N] + (1,2,\ldots,m)(m+1,\ldots,N).\end{aligned}$$

Hence the result.

Lemma 9.

$$\mathfrak{a} = \sum_{\mathfrak{a}_1 \circ \ldots \circ \mathfrak{a}_K = \mathfrak{a}} [\mathfrak{a}_1] \ldots [\mathfrak{a}_K].$$

Proof. We proceed again by induction with respect to the length of $\mathfrak{a}$. For $|\mathfrak{a}| = 1$ is $[k] = k$ and all is clear. The induction from length $= N-1$ to length $= N$ is performed by the formulas of lemma 7.

Let J be a complex algebra with unit element 1 and inversion; i.e., there exists a mapping $x \mapsto x^*$ from J into J such that

$$\begin{aligned}(x^*)^* &= 1\\ 1^* &= 1\\ (x+y)^* &= x^* + y^*\\ (\lambda x)^* &= \bar{\lambda} x^*\end{aligned}$$

for $x, y \in J$, $\lambda \in \mathbb{C}$.

An element x is called unitary, if $xx^* = x^*x = 1$.

Let α be a linear mapping from $\mathcal{F}$ into J, with the properties

$$\alpha\mathfrak{m} \text{ is unitary}$$

$$\alpha\mathfrak{m}_1\mathfrak{m}_2 = \alpha\mathfrak{m}_2\alpha\mathfrak{m}_1$$

if $\mathfrak{m}, \mathfrak{m}_1, \mathfrak{m}_2$ are monomials. The second property implies that α is an antihomomorphism.

Let β be a second linear mapping from $\mathcal{F}$ into J with the property

$$\beta\mathfrak{m}_1\mathfrak{m}_2 = \beta\mathfrak{m}_2\alpha\mathfrak{m}_1 + \beta\mathfrak{m}_1$$

for monomials $\mathfrak{m}_1, \mathfrak{m}_2$. For arbitrary elements f_1 and f_2 this implies

$$\beta f_1 f_2 = \beta f_2 \alpha f_1 + \sigma f_2 \beta f_1$$

where σ is the homomorphism from $\mathcal{F}$ into $\mathbb{R}$ given by

$$\sigma(\sum c_i\mathfrak{m}_i) = \sum c_i$$

if $\mathfrak{m}_i$ are monomials. β is an antiderivation from $\mathcal{F}$ into J.

Define for monomials $\mathfrak{m}$

$$\tilde{\beta}\mathfrak{m} = \alpha\mathfrak{m}(\beta\mathfrak{m})^*$$

and extend $\tilde{\beta}$ to $\mathcal{F}$ as a linear mapping.

Then

$$\tilde{\beta}\mathfrak{m}_1\mathfrak{m}_2 = \alpha(\mathfrak{m}_1\mathfrak{m}_2)\beta(\mathfrak{m}_1\mathfrak{m}_2)^*$$
$$= \alpha\mathfrak{m}_2\alpha\mathfrak{m}_1(\alpha\mathfrak{m}_1^*\beta\mathfrak{m}_2^* + \beta\mathfrak{m}_1^*)$$

and so

$$\tilde{\beta}\mathfrak{m}_1\mathfrak{m}_2 = \alpha\mathfrak{m}_2\tilde{\beta}\mathfrak{m}_1 + \tilde{\beta}\mathfrak{m}_2$$
$$\tilde{\beta}f_1f_2 = \alpha f_2\tilde{\beta}f_1 + \sigma f_1\tilde{\beta}f_2$$

Finally define for monomials

$$\tilde{\tilde{\beta}}\mathfrak{m} = \beta\mathfrak{m}\cdot\beta\mathfrak{m}^* = \tilde{\beta}\mathfrak{m}^*\cdot\bar{\beta}\mathfrak{m}$$

and extend it to $\mathfrak{f}$ in a linear way. Then

$$\tilde{\tilde{\beta}}\mathfrak{m}_1\mathfrak{m}_2 = \tilde{\tilde{\beta}}\mathfrak{m}_1 + \tilde{\tilde{\beta}}\mathfrak{m}_2 + \beta\mathfrak{m}_2\tilde{\beta}\mathfrak{m}_1 + \tilde{\beta}\mathfrak{m}_1^{*}\ \beta\mathfrak{m}_2^{*}$$

$$\tilde{\tilde{\beta}}f_1f_2 = \sigma f_2\tilde{\tilde{\beta}}f_1 + \sigma f_1\tilde{\tilde{\beta}}f_2 + \beta f_2\cdot\tilde{\beta}f_1 + \tilde{\beta}f_1^{*}\cdot\beta f_2^{*}\ ;$$

<u>Lemma 10.</u> The following formulae hold

$$\alpha(1,2,\dots,N) = \sum_{\mathfrak{a}_1\circ\dots\circ\mathfrak{a}_K=(1,2,\dots,N)} \alpha[\mathfrak{a}_K]\,\alpha[\mathfrak{a}_{K-1}]\cdots\alpha[\mathfrak{a}_1]$$

$$\beta(1,2,\dots,N) = \sum_{k=1}^{N}\ \sum_{\mathfrak{a}_1\circ\dots\circ\mathfrak{a}_K=(1,\dots,k)} \beta[\mathfrak{a}_K]\,\alpha[\mathfrak{a}_{K-1}]\cdots\alpha[\mathfrak{a}_1]$$

$$\tilde{\beta}(1,2,\dots,N) = \sum_{k=1}^{N}\ \sum_{\mathfrak{a}_1\circ\dots\circ\mathfrak{a}_K=(k,\dots,N)} \alpha[\mathfrak{a}_K]\cdots\alpha[\mathfrak{a}_2]\,\tilde{\beta}[\mathfrak{a}_1]$$

$$\tilde{\tilde{\beta}}(1,2,\dots,N) = \sum_{1\leqslant k\leqslant \ell\leqslant N}\Bigg(\tilde{\tilde{\beta}}[k,\dots,\ell] + \sum_{\mathfrak{a}_1\circ\dots\circ\mathfrak{a}_K=(k,\dots,\ell)}$$

$$\Big(\beta[\mathfrak{a}_K]\,\alpha[\mathfrak{a}_{K-1}]\cdots\alpha[\mathfrak{a}_2]\,\tilde{\beta}[\mathfrak{a}_1] + \tilde{\beta}[\mathfrak{a}_1]^{*}\,\alpha[\mathfrak{a}_2]^{*}\cdots$$
$$\cdots\alpha[\mathfrak{a}_{K-1}]^{*}\,\beta[\mathfrak{a}_K]^{*}\Big)\Bigg)$$

<u>Proof.</u> At first we establish the formula

$$\sigma[\mathfrak{a}] = \begin{cases} 1 & \text{for } |\mathfrak{a}| = 1 \\ \sigma & \text{for } |\mathfrak{a}| > 1 \end{cases}$$

by induction with respect to the length $|\mathfrak{a}|$. For $|\mathfrak{a}| = 1$ the result is clear. Assume validity for $|\mathfrak{a}| \leqslant N-1$. Then lemma 7 gives

$$1 = \sigma(1,2,\dots,N) = \sigma[1,2,\dots,N] + \sigma(1,2,\dots,N-1)\,\sigma[N]$$
$$= \sigma[1,2,\dots,N] + 1.$$

The first formula of lemma 10 follows directly from lemma 9 and from the fact that α is an antihomomorphism. In order to prove the second formula use the first formula of lemma 7 and get

$$\beta(1,2,\dots,N) = \beta[1,2,\dots,N] + \beta[2,\dots,N]\,\alpha(1)$$
$$+\ \beta[3,\dots,N]\,\alpha(1,2) + \dots + \beta[N]\,\alpha(1,2,\dots,N-1)$$

$$+ \ \beta(1,2,\ldots, N-1)$$

$$= \sum_{\mathfrak{a}_1 \circ \ldots \circ \mathfrak{a}_K = (1,2,\ldots,N)} \beta[\mathfrak{a}_K]\,\alpha[\mathfrak{a}_{K-1}] \ldots \alpha[\mathfrak{a}_1]$$

$$+ \ \beta(1,2,\ldots, N-1).$$

Using induction one proves the second formula. The third formula is shown in a similar way by applying the second formula of lemma 7. In a similar way, by using the three precedent formulas, one obtains the fourth formula.

II.3. The second formula

In order to apply the results of II.2. we extend at first the functions $\alpha(\mathfrak{a}) = \alpha\mathfrak{a}$, $\beta\mathfrak{a}$, $\tilde{\beta}\mathfrak{a}$, $\tilde{\tilde{\beta}}\mathfrak{a}$ as defined in II.1. to the whole of $\mathfrak{F}$ and set for a monomial $\mathfrak{m} = \mathfrak{a}_1 \ldots \mathfrak{a}_K$

$$\alpha\mathfrak{m} = \alpha\mathfrak{a}_K * \alpha\mathfrak{a}_{K-1} * \ldots * \alpha\mathfrak{a}_1$$

$$\beta\mathfrak{m} = \beta\mathfrak{a}_K * \alpha\mathfrak{a}_{K-1} * \ldots * \alpha\mathfrak{a}_1 + \beta\mathfrak{a}_{K-1} * \alpha\mathfrak{a}_{K-2} * \ldots * \alpha\mathfrak{a}_1 + \ldots + \beta\mathfrak{a}_1$$

$$\tilde{\beta}\mathfrak{m} = \alpha\mathfrak{m} * \beta\mathfrak{m}^+$$

$$\tilde{\tilde{\beta}}\mathfrak{m} = \beta\mathfrak{m} * \beta\mathfrak{m}^*.$$

Then α, β, $\tilde{\beta}$, $\tilde{\tilde{\beta}}$ have all the properties of the quantities denoted in II.2. by the same letters. An easy application of lemma 1 yields

Lemma 11. Let $u_m > \tau$ then for $f \in \mathfrak{F}$

$$\alpha f = \alpha\gamma_m f$$

$$\beta f = \beta\gamma_m f$$

$$\tilde{\beta} f = \tilde{\beta}\gamma_m f$$

$$\tilde{\tilde{\beta}} f = \tilde{\tilde{\beta}}\gamma_m f$$

Corollary. If $u_m > \tau$ for $1 \leqslant m < N$

$$\alpha[1, 2, \ldots, N] = 0$$
$$\beta[1, 2, \ldots, N] = 0$$
$$\tilde{\beta}[1, 2, \ldots, N] = 0$$
$$\tilde{\tilde{\beta}}[1, 2, \ldots, N] = 0$$

Proof. If $u_m > \tau$, then by lemma 11 e.g.

$\alpha[1, 2, \ldots, N] = \alpha \varphi_m[1, 2, \ldots, N] = 0$ using lemma 8.

Theorem 2. Define

$$\alpha_N = E\alpha[1, 2, \ldots, N]$$
$$\beta_N = E\beta[1, 2, \ldots, N]$$
$$\tilde{\beta}_N = E\tilde{\beta}[1, 2, \ldots, N]$$
$$\tilde{\tilde{\beta}}_N = E\tilde{\tilde{\beta}}[1, 2, \ldots, N]$$

Then α_N is a bounded $\mathcal{L}(\mathcal{H})$-valued measure on $\mathbb{R}_+$, β_N and $\tilde{\beta}_N$ are bounded borelian functions with support in $[-\tau/2, \infty[$, β_N is a continuous $\mathcal{L}(\mathcal{H})$-valued function on $\mathbb{R}$ vanishing at infinity.

If $p\{u > \tau\} > \frac{1}{2}$, then there exist the sums

$$\alpha = \sum_{N=1}^{\infty} \alpha_N$$

$$\beta = \sum_{N=1}^{\infty} \beta_N$$

$$\tilde{\beta} = \sum_{N=1}^{\infty} \tilde{\beta}_N$$

$$\tilde{\tilde{\beta}} = \sum_{N=1}^{\infty} \tilde{\tilde{\beta}}_N .$$

The first sum converges in the norm

$$\mu \longmapsto \|\mu\| = \sup\{|\langle \mu, \psi\rangle| : \psi \in C_*(\mathbb{R}_+), |\psi| \leqslant 1\}$$

on the space of all $\mathcal{L}(\mathcal{H})$-valued measures on $\mathbb{R}_+$. The other three sums converge uniformly. So β and $\tilde{\beta}$ are bounded borelian functions on $[-\tau/2, \infty[$ and $\tilde{\tilde{\beta}}$ is continuous on $\mathbb{R}$ and vanishes at infinity.

Assume $p\{u = 0\} < \frac{1}{3}$. Then the measure has an inverse in the algebra of all $\mathcal{L}(\mathcal{H})$-valued measures on $\mathbb{R}_+$.

Finally we get the formula

$$R = \frac{1}{Eu}(\tilde{\tilde{\beta}} + \beta * (\delta - \alpha)^{*-1} * \tilde{\beta}$$
$$+ \tilde{\beta}^{+} * (\delta - \alpha^{+})^{*-1} * \beta^{*}).$$

Proof. The statements about α_N, β_N, $\tilde{\beta}_N$, $\tilde{\tilde{\beta}}_N$ are obvious. At first we prove the convergence of α. By corollary of lemma 11 we have

$$\alpha_N = E\alpha[1,2,\ldots, N]\, \mathbb{1}\{u_1 \leqslant \tau\} \ldots \mathbb{1}\{u_{N-1} \leqslant \tau\}$$
$$= \sum_{\mathfrak{a}_1 \circ \ldots \circ \mathfrak{a}_K = (1,2,\ldots,N)} (-1)^{K+1}\, \alpha(\mathfrak{a}_1 \ldots \mathfrak{a}_K)\, \mathbb{1}\{u_1 \leqslant \tau\} \ldots \mathbb{1}\{u_{N-1} \leqslant \tau\}.$$

As $\alpha(\mathfrak{a}_1 \ldots \mathfrak{a}_K)$ has norm $= 1$ and as there are 2^{N-1} terms in the sum we get

$$\|\alpha_N\| \leqslant 2^{N-1}\, p\{u \leqslant \tau\}^{N-1}.$$

Call

$$b = p\{u \leqslant \tau\}.$$

By assumption $b < \frac{1}{2}$. Therefore

$$\|\alpha_N\| \leqslant (2b)^{N-1}$$

and

$$\sum_{N=1}^{\infty} \|\alpha_N\| \leqslant \frac{1}{1-2b} < \infty.$$

We investigate now the behaviour of β_N. By the corollary of lemma 11

$$\beta_N = E\beta[1,2,\ldots,N]\, \mathbb{1}\{u_1 \leqslant \tau\} \ldots \mathbb{1}\{u_{N-1} \leqslant \tau\}$$
$$= \sum_{K=1}^{N} (-1)^{K+1} \sum_{\mathfrak{a}_1 \circ \ldots \circ \mathfrak{a}_K = (1,2,\ldots,N)} E\beta(\mathfrak{a}_1 \ldots \mathfrak{a}_K)\mathbb{1}\{u_1 \leqslant \tau\} \ldots \mathbb{1}\{u_{N-1} \leqslant \tau\}$$

and

$$\beta(\sigma_1 \ldots \sigma_K) = \beta\sigma_K * \alpha\sigma_{K-1} * \cdots * \alpha\sigma_1$$
$$+ \beta\sigma_{K-1} * \alpha\sigma_{K-2} * \cdots * \alpha\sigma_1$$
$$+ \ldots + \beta\sigma_1 .$$

Now $\beta\sigma_j$ has its support in $[-\tau/2, \infty[$ and $\alpha\sigma_j$ its support in $\mathbb{R}_+$. Therefore $\beta(\sigma_1 \ldots \sigma_K)$ has its support in $[-\tau/2, \infty[$. As $\beta\sigma_j$ is bounded by 2 and $\alpha\sigma_j$ has a norm $\leqslant 1$, the function $\beta\sigma_j * \alpha\sigma_{j-1} * \cdots * \alpha\sigma_1$ is bounded by 2 and $\beta(\sigma_1 \ldots \sigma_K)$ is bounded by $2K \leqslant 2N$. So

$$|\beta(\sigma_1 \ldots \sigma_K)| \leqslant 2N\, \mathbb{1}[-\tau/2, \infty[$$

and

$$|\beta_N| \leqslant 2N\, 2^{N-1}\, b^{N-1}\, \mathbb{1}[-\tau/2, \infty[$$

Again

$$\sum_{N=1}^{\infty} |\beta_N| < \infty .$$

In the same way one proves that

$$\sum_{N=1}^{\infty} |\tilde{\beta}_N| < \infty .$$

We have

$$\tilde{\tilde{\beta}}_N = E\, \tilde{\tilde{\beta}}[1,2,\ldots,N]\, \mathbb{1}\{u_1 \leqslant \tau\} \ldots \mathbb{1}\{u_{N-1} \leqslant \tau\}$$

$$= \sum_{\sigma_1 \circ \cdots \circ \sigma_K = (1,2,\ldots,N)} (-1)^K \tilde{\tilde{\beta}}(\sigma_1 \ldots \sigma_K)\, \mathbb{1}\{u_1 \leqslant \tau\} \ldots \mathbb{1}\{u_{N-1} \leqslant \tau\}$$

As

$$\tilde{\tilde{\beta}}\sigma_1 \ldots \sigma_K = \sum_{i=1}^{K} \tilde{\tilde{\beta}}\sigma_i + \Big[\sum_{1 \leqslant i < j \leqslant K} \beta\sigma_j * \alpha\sigma_{j-1} * \cdots$$
$$\cdots * \alpha\sigma_{i+1} * \tilde{\beta}\sigma_i \Big] + \Big[\qquad \Big]^+$$

and

$$|\beta \alpha_j| \leqslant 2\, \mathbb{1}[-\tau/2, U(\alpha_j) + \tau/2],$$

we get

$$|\tilde{\tilde{\beta}} \alpha_j| \leqslant 4\,(\tau + U(\alpha_j))$$

$$|\beta \alpha_j * \alpha \alpha_{j-1} * \ldots * \tilde{\beta} \alpha_i| \leqslant 4(\tau + U(\alpha_i)) \leqslant 4(\tau + U(\alpha_i) + U(\alpha_j))$$

and

$$|\tilde{\tilde{\beta}} \alpha_1 \ldots \alpha_K| \leqslant 4 \sum_{i=1}^{K} (U(\alpha_i) + \tau) + 8 \sum_{1 \leqslant i < j \leqslant K} (U(\alpha_i) + U(\alpha_j) + \tau)$$

$$\leqslant 4\, N^2 (u_1 + \ldots u_N + \tau).$$

Finally

$$|\tilde{\tilde{\beta}}_N| \leqslant 4N^2\, 2^{N-1}\, E(u_1 + \ldots + u_N + \tau)\, \mathbb{1}\{u_1 \leqslant \tau\} \ldots \mathbb{1}\{u_{N-1} \leqslant \tau\}$$

$$\leqslant 4N^2\, 2^{N-1} ((N-2)\, b^{N-2}\, Eu\, \mathbb{1}\{u \leqslant \tau\} + a^{N-1}(Eu + \tau))$$

and

$$\sum_{N=1}^{\infty} |\tilde{\tilde{\beta}}_N| < \infty .$$

So for $p\{u > \tau\} < \frac{1}{2}$ the convergence of α, β, $\tilde{\beta}$, $\tilde{\tilde{\beta}}$ in the sense described in the theorem is proved. In order to prove the existence of

$$(\delta - \alpha)^{*-1} = \delta + \alpha + \alpha^{*2} + \ldots$$

we have to apply lemma 6 and to check whether $|\alpha|\{0\} < 1$. As

$$(\mu * \nu)\{0\} = \mu\{0\}\nu\{0\}$$

for measures on $\mathbb{R}_+$ we get

$$|E\, \alpha \alpha_K * \ldots * \alpha \alpha_1 \{0\}|$$

$$\leqslant E\, \delta_{u_1 + u_N} \{0\}$$

$$= p\{u=0\}^N.$$

Hence

$$|\alpha|\{0\} \leqslant \sum_{N=1}^{\infty} p\{u=0\}^{N}\, 2^{N-1} = \frac{p\{u=0\}}{1-2p\{u=0\}} < 1,$$

because $p\{u=0\} < \frac{1}{3}$ by assumption.

In order to prove the final formula observe that

$$W_N * W_N^+ = \tilde{\tilde{\beta}}(1, 2, \ldots, N)$$

and apply lemmata 10 and 5 with the explications made after the proof of lemma 5. Then

$$\frac{1}{N} W_N * W_N^+ = \frac{1}{N} \sum_{1 \leqslant k \leqslant \ell \leqslant N} \Bigg\{ \tilde{\tilde{\beta}}_{\ell-k+1}$$

$$+ \Bigg[\sum_{\substack{n_1+\ldots+n_K=\ell-k+1 \\ n_i \geqslant 1,\ K \geqslant 2}} \beta_{n_K} * \alpha_{n_{K-1}} * \ldots * \alpha_{n_2} * \tilde{\beta}_{n_1} \Bigg] + \Bigg[\quad \Bigg]^+ \Bigg\}$$

$$= \sum_{n=1}^{N} \frac{N-n+1}{N} \Bigg\{ \tilde{\tilde{\beta}}_n +$$

$$+ \Bigg[\sum_{\substack{n_1+\ldots+n_K=n \\ n_i \geqslant 1,\ K \geqslant 2}} \beta_{n_K} * \alpha_{n_{K-1}} * \ldots * \alpha_{n_2} * \tilde{\beta}_{n_1} \Bigg] + \Bigg[\quad \Bigg]^+ \Bigg\}.$$

Going formally with N to infinity we obtain

$$\sum_{n=1}^{\infty} \tilde{\tilde{\beta}}_n + \Bigg[\sum_{K=2}^{\infty} \sum_{n_1,\ldots,n_{K-1}}^{\infty} \beta_{n_K} * \alpha_{n_{K-1}} \ldots * \alpha_{n_2} * \tilde{\beta}_{n_1} \Bigg] + \Bigg[\quad \Bigg]^+ =$$

$$= \tilde{\tilde{\beta}} + \Bigg[\sum_{K=2}^{\infty} \beta * \alpha^{*(K-2)} * \tilde{\beta} \Bigg] + \Bigg[\quad \Bigg]^+$$

$$= \tilde{\tilde{\beta}} + \beta * (\delta-\alpha)^{*-1} * \tilde{\beta} + \tilde{\beta}^+ * (\delta-\alpha^+)^{*-1} * \beta^+.$$

In order to justify this formal procedure it is sufficient to check the absolute convergence of all the series involved in the the formulae. This can easily be done by the estimates given above.

II.4. Specialization of the Probability Measure

Assume now that the measure p on $Z \times \mathbb{R}_+$ is of the form $p = p_Z \otimes p_c$ as described at the beginning of II.1. In this case simpler forms can be given to the first and the second formula.

We begin with the first formula. Recall

$U(\sigma) = u_1 + \ldots + u_{K-1} + u_K$, if K is the first j such that $u_j > \tau$. Put

$$Y(\sigma) = u_1 + \ldots + u_{K-1} + \tau$$

$$y(\sigma) = u_K - \tau.$$

Then $y(\sigma)$ is distributed with respect to the law p_c, as can easily be checked. The function

$$\Phi(\sigma) = h(\varphi(\zeta_1) + \varphi(\zeta_2)[u_1] + \ldots + \varphi(\zeta_k)[u_1 + \ldots + u_{K-1}])$$

has its support in $[-\tau/2, Y(\sigma)-\tau/2]$. Define

$$H = E \int_{-\infty}^{+\infty} \Phi(\sigma)\delta_{Y(\sigma)}$$

$$\theta = E\left(\int_{-\infty}^{t} \Phi(\sigma) - \mathbb{1}_{\leqslant 0}(t) - \mathbb{1}_{>Y(\sigma)} \int_{-\infty}^{+\infty} \Phi(\sigma)\right)$$

$$\tilde{\theta} = E\left(\int_{Y(\sigma)-t}^{\infty} \Phi(\sigma) - \mathbb{1}_{<0}(t) - \mathbb{1}_{\geqslant Y(\sigma)} \int_{-\infty}^{+\infty} \Phi(\sigma)\right)$$

$$\tilde{\tilde{\theta}} = E\left(\int_{-\infty}^{t} \Phi(\sigma) - \mathbb{1}_{\leqslant 0}(t) - \mathbb{1}_{>Y(\sigma)} \int_{-\infty}^{+\infty} \Phi(\sigma)\right) *$$

$$\left(\int_{-\infty}^{t} \Phi(\sigma) - \mathbb{1}_{\leqslant 0}(t) - \mathbb{1}_{>Y(\sigma)} \int_{-\infty}^{+\infty} \Phi(\sigma)\right)^{+}.$$

Then we have

<u>Theorem 3.</u> For R the formula holds

$$R = ce^{-c\tau}\tilde{\tilde{\theta}} + e^{-c\tau}(\delta+c\theta) * (\delta'+c\delta-cH)^{*-1} * (\delta+c\tilde{\theta}) +$$

$$+ e^{-c\tau}(\delta+c\bar{\theta}^{+}) * [(\delta'+c\delta-cH)^{*-1}] * (\delta+c\theta^{+})$$

where the inverse $(\delta'+c\delta-cH)^{*-1}$ is to be taken in the sense of the convolution algebra of Schwartz distribution on $\mathbb{R}_+$, $\delta' = \frac{d}{dt}\delta$.

<u>Proof.</u> We have $p\{u>\tau\}/Eu = ce^{-c\tau}$. Call

$$\pi = \pi(\sigma) = \overset{+\infty}{\underset{-\infty}{\tilde{\omega}}} \Phi(\sigma)\delta_{Y(\sigma)}$$

$$\rho = \rho(\sigma) = \overset{t}{\underset{-\infty}{\tilde{\omega}}} \Phi(\sigma) - \mathbb{1}_{\leq 0}(t) - \mathbb{1}_{>Y(\sigma)} \overset{+\infty}{\underset{-\infty}{\tilde{\omega}}} \Phi(\sigma)$$

$$\tilde{\rho} = \pi * \rho^+, \ \tilde{\tilde{\rho}} = \rho * \rho^+.$$

Then

$$\alpha(\sigma) = \pi * \delta_{y(\sigma)}$$

$$\beta(\sigma) = \rho + \mathbb{1}[Y(\sigma), U(\sigma)[\overset{+\infty}{\underset{-\infty}{\tilde{\omega}}} \Phi(\sigma)$$

$$= \rho + \pi * \mathbb{1}[0, y(\sigma)[$$

$$\tilde{\beta}(\sigma) = \alpha(\sigma) * \beta(\sigma)^+$$

$$= \tilde{\rho} * \delta_{y(\sigma)} + \mathbb{1}]0, y(\sigma)]$$

$$\tilde{\tilde{\beta}}(\sigma) = \rho * \rho^+ + \rho * \pi^+ * \mathbb{1}[0, y(\sigma)[^+$$

$$+ \pi * \rho^+ * \mathbb{1}[0, y(\sigma)[+ \mathbb{1}[0, y(\sigma)[* \mathbb{1}[0, y(\sigma)[^+$$

$$= \tilde{\tilde{\rho}} + \tilde{\rho}^+ * \mathbb{1}[0, y(\sigma)[^+ + \tilde{\rho} * \mathbb{1}[0, y(\sigma)[$$

$$+ \mathbb{1}[-y(\sigma), 0] \cdot (y(\sigma)+t) + \mathbb{1}[0, y(\sigma)] \big((y(\sigma)-t)\big).$$

We have

$$H = E\pi$$

$$\theta = E\rho$$

$$\tilde{\theta} = E\tilde{\rho}$$

$$\tilde{\tilde{\theta}} = E\tilde{\tilde{\rho}}$$

and get

$$\eta = E\alpha(\sigma) = H * p_c$$

$$\vartheta = E\beta(\sigma) = \theta + H * E\, \mathbb{1}[0, y(\sigma)]$$

$$\tilde{\vartheta} = E\tilde{\beta}(\sigma) = \tilde{\theta} * p_c + \mathbb{1}[0, y(\sigma)]$$

$$\tilde{\tilde{\vartheta}} = E\tilde{\tilde{\beta}}(\sigma) = \tilde{\tilde{\theta}} + \tilde{\theta}^+ * E\, \mathbb{1}[-y(\sigma), 0]$$

$$+ \tilde{\theta} * E\, \mathbb{1}[0, y(\sigma)] + E\, \mathbb{1}[0, y(\sigma)]\, (y(\sigma)-t)$$

$$+ E\,\mathbb{1}[-y(\sigma), 0]\,(y(\sigma)+t).$$

We get for R

$$R = ce^{-c\tau}(\tilde{\tilde{\vartheta}} + [\vartheta * (\delta-n)^{*-1} * \tilde{\vartheta}] + [\quad]^{+})$$

$$= ce^{-c\tau}(\tilde{\tilde{\theta}} + [+\tilde{\theta} * E\,\mathbb{1}[0, y(\sigma)] + E\,\mathbb{1}[0, y(\sigma)] \cdot$$

$$\cdot (y(\sigma)-t) + \vartheta * \frac{\delta}{\delta-n} * \tilde{\vartheta}] + [\quad]^{+}).$$

In order to calculate $[\quad]$ we use the Heaviside-calculus

$$p_c = \frac{c\delta}{\delta'+c\delta}$$

$$E\,\mathbb{1}[0, y(\sigma)] = \frac{\delta}{\delta'+c\delta}$$

$$E\,\mathbb{1}[0, y(\sigma)]\,(y(\sigma)-t) = \frac{1}{c}\,\frac{\delta}{\delta'+c\delta}.$$

Then

$$[\quad] = \tilde{\theta} * \frac{\delta}{\delta'+c\delta} + \frac{1}{c}\,\frac{\delta}{\delta'+c\delta}$$

$$+ \left(\theta + H * \frac{\delta}{\delta'+c\delta}\right) * \frac{\delta}{\delta - H * \frac{c\delta}{\delta'+c\delta}} * \left(\tilde{\theta} * \frac{c\delta}{\delta'+c\delta} + \frac{\delta}{\delta'+c\delta}\right)$$

$$= \left[\left(\theta + H * \frac{\delta}{\delta'+c\delta}\right) * \frac{\delta}{\delta - H * \frac{c\delta}{\delta'+c\delta}} + \frac{\delta}{c}\right] * \left(\tilde{\theta} * \frac{c\delta}{\delta'+c\delta} + \frac{\delta}{\delta'+c\delta}\right)$$

$$= \left(\theta + H * \frac{\delta}{\delta'+c\delta} + \frac{\delta}{c} - H * \frac{\delta}{\delta'+c\delta}\right) * \frac{\delta}{\delta - \frac{cH}{\delta'+c\delta}} * \left(\frac{c\tilde{\theta}}{\delta'+c\delta} + \frac{\delta}{\delta'+c\delta}\right)$$

$$= \frac{1}{c}\,(\delta+c\theta) * \frac{\delta}{\delta'+c\delta-cH} * (\delta+c\tilde{\theta}).$$

The second formula gets into a similar shape. Let $\mathfrak{a} = (k, \ldots, \ell)$ be an interval. Then call

$$V(\mathfrak{a}) = u_k + \ldots + u_{\ell-1}$$

$$A\mathfrak{a} = \delta_{V(\mathfrak{a})} \overset{+\infty}{\underset{-\infty}{\tilde{\omega}}} \Phi(\mathfrak{a})$$

$$B\mathfrak{a} = \overset{t}{\underset{-\infty}{\tilde{\omega}}} \Phi(\mathfrak{a}) - \mathbb{1}_{\leq 0} - \mathbb{1}_{> V(\mathfrak{a})} \overset{+\infty}{\underset{-\infty}{\tilde{\omega}}} \Phi(\mathfrak{a}).$$

For monomials $\mathfrak{m} = \mathfrak{a}_1 \ldots \mathfrak{a}_K$

$$A\mathfrak{m} = A\mathfrak{a}_K * \alpha\mathfrak{a}_{K-1} * \ldots * \alpha\mathfrak{a}_1$$

$$B\mathfrak{m} = B\mathfrak{a}_K * \alpha\mathfrak{a}_{K-1} * \ldots * \alpha\mathfrak{a}_1 + \beta\mathfrak{a}_{K-1} * \alpha\mathfrak{a}_{K-2} * \ldots * \alpha\mathfrak{a}_1 + \ldots + \beta\mathfrak{a}_1$$

$$\tilde{B}\mathfrak{m} = A\mathfrak{m} * B\mathfrak{m}^+$$

$$\tilde{\tilde{B}}\mathfrak{m} = B\mathfrak{m} * B\mathfrak{m}^+ .$$

We define

$$A_N = EA[1, 2, \ldots, N]$$

$$B_N = EB[1, 2, \ldots, N]$$

$$\tilde{B}_N = E\tilde{B}[1, 2, \ldots, N]$$

$$\tilde{\tilde{B}}_N = E\tilde{\tilde{B}}[1, 2, \ldots, N]$$

By similar reasoning as just below we get (for notation compare II.3.)

Theorem 4. Let $c\tau < \log 2$. Then the sums converge uniformly

$$B = \sum_{N=1}^{\infty} B_N$$

$$\tilde{B} = \sum_{N=1}^{\infty} \tilde{B}_N$$

$$\tilde{\tilde{B}} = \sum_{N=1}^{\infty} \tilde{\tilde{B}}_N .$$

The functions B and $\tilde{B}$ are borelian and bounded with support in $[-\tau/2,\infty]$. The functionn $\tilde{\tilde{B}}$ is continuous and vanishes at infinity. The sum

$$A = \sum_{N=1} A_N$$

converges in the norm $\|\cdot\|$ to a measure A on $\mathbb{R}_+$.

We have the formula

$$R = c\tilde{B} + (\delta + cB) * (\delta' + c\delta - cA)^{*-1} * (\delta + c\tilde{B})$$

$$+ (\delta + c\tilde{B}^{+}) * \left[(\delta' + c\delta - cA)^{*-1}\right]^{+} * (\delta + cB^{+}).$$

This formula has been discussed in I.

II.5. The formulae for $-\frac{d^2}{dt^2} R$

Without many proofs we list the formulae corresponding to theorem 1 to 4 for $-\frac{d^2}{dt^2} R$ or after Fourier transform for $\omega^2 I(\omega)$. These formulae seem to be useful on the line-wings (cf. I).

Let $\mathfrak{n} \subset \mathbb{N}$ be an interval and put

$$\gamma\mathfrak{n}(t) = \Phi(\mathfrak{n})\,(t)\ \underset{-\infty}{\overset{t}{\varpi}}\ \Phi(\mathfrak{n})$$

which is a function with support $\subset [-\tau/2,\ U(\mathfrak{n}) + \tau/2]$. Then

$$\beta\mathfrak{n} = \underset{-\infty}{\overset{t}{\varpi}}\ \Phi(\mathfrak{n}) - 1\!\!1_{\leqslant 0} - 1\!\!1_{>U(\mathfrak{n})}\ \underset{-\infty}{\overset{+\infty}{\varpi}}\ \Phi(\mathfrak{n})$$

and

$$\partial\beta\mathfrak{n} = \gamma\mathfrak{n} + \delta - \alpha\mathfrak{n}$$

with

$$\partial = \frac{1}{i}\,\frac{d}{dt}\,.$$

Define

$$\tilde{\gamma}\mathfrak{n} = \alpha\mathfrak{n} * \gamma\mathfrak{n}^{+}$$

$$\tilde{\tilde{\gamma}}\mathfrak{n} = \gamma\mathfrak{n} * \gamma\mathfrak{n}^{+}.$$

Then

$$\partial\tilde{\beta}\mathfrak{n} = \tilde{\gamma}\mathfrak{n} - \delta + \alpha\mathfrak{n}$$

$$\partial^2\tilde{\tilde{\beta}}\mathfrak{n} = \tilde{\tilde{\gamma}}\mathfrak{n} + \gamma\mathfrak{n} + \gamma\mathfrak{n}^{+} - \tilde{\gamma}\mathfrak{n} - \tilde{\gamma}\mathfrak{n}^{+} + 2\delta - \alpha\mathfrak{n} - \alpha\mathfrak{n}^{+}.$$

Define now with the notation of II.1.

$$\varkappa = E\gamma(\sigma)$$

$$\tilde{\varkappa} = E\tilde{\gamma}(\sigma)$$

$$\tilde{\tilde{\varkappa}} = E\tilde{\tilde{\gamma}}(\sigma).$$

Then we get

$$\partial\vartheta = \varkappa + \delta - \eta$$

$$\partial\tilde{\vartheta} = \tilde{\varkappa} - \delta + \eta$$

$$\partial^2\tilde{\tilde{\vartheta}} = \tilde{\tilde{\varkappa}} + \varkappa + \varkappa^+ - \tilde{\varkappa} - \tilde{\varkappa}^+ + 2\delta - \eta - \eta^+$$

hence (cf. Theorem 1)

$$\partial^2 R = \frac{p\{u > \tau\}}{Eu}\left(\tilde{\tilde{\varkappa}} + \left[\varkappa - \tilde{\varkappa} + \delta - \eta + (\varkappa + \delta - \eta) * \frac{\delta}{\delta - \eta} * (\tilde{\varkappa} - \delta + \eta)\right] + \left[\ \right]^+\right).$$

An easy calculation leads to the analogue of theorem 1:

$$(1)\quad \partial^2 R = \frac{p\{u > \tau\}}{Eu}\left(\tilde{\tilde{\varkappa}} + \varkappa * \frac{\delta}{\delta - \eta} * \tilde{\varkappa} + \tilde{\varkappa}^+ * \frac{\delta}{\delta - \eta^+} * \varkappa^+\right).$$

Extend γ to monomials $\mathfrak{m} = \mathfrak{a}_1 \dots \mathfrak{a}_K$ by

$$\gamma\mathfrak{m} = \gamma\mathfrak{a}_K * \alpha\mathfrak{a}_{K-1} * \dots * \alpha\mathfrak{a}_1$$
$$+ \gamma\mathfrak{a}_{K-1} * \alpha\mathfrak{a}_{K-2} * \dots * \alpha\mathfrak{a}_1$$
$$+ \dots + \gamma\mathfrak{a}_1$$

define $\tilde{\gamma}\mathfrak{m} = \alpha\mathfrak{m} * \gamma\mathfrak{m}^+$ and $\tilde{\tilde{\gamma}}\mathfrak{m} = \gamma\mathfrak{m} * \gamma\mathfrak{m}^+$.

The relations combining $\beta\mathfrak{a}$, $\tilde{\beta}\mathfrak{a}$, $\tilde{\tilde{\beta}}\mathfrak{a}$ and $\gamma\mathfrak{a}$, $\tilde{\gamma}\mathfrak{a}$, $\tilde{\tilde{\gamma}}\mathfrak{a}$ remain true, if we replace $\mathfrak{a}$ by $\mathfrak{m}$. Put

$$\gamma_N = E\gamma[1, 2, \dots, N]$$

$$\tilde{\gamma}_N = E\tilde{\gamma}[1, 2, \dots, N]$$

$$\tilde{\tilde{\gamma}}_N = E\tilde{\tilde{\gamma}}[1, 2, \dots, N]$$

and

$$\gamma = \sum_{N=1}^{\infty} \gamma_N, \quad \tilde{\gamma} = \sum_{N=1}^{\infty} \tilde{\gamma}_N, \quad \tilde{\tilde{\gamma}} = \sum_{N=1}^{\infty} \tilde{\tilde{\gamma}}_N.$$

Then we get (cf. theorem 2):

(2) $\partial^2 R = \frac{1}{Eu}\left(\tilde{\tilde{\gamma}} + \gamma * \frac{\delta}{\delta-\alpha} * \tilde{\gamma} + \tilde{\gamma}^+ * \frac{\delta}{\delta-\alpha^+} * \gamma^+\right)$

for $p\{u > \tau\} > \frac{1}{2}$ and $p\{u = 0\} < \frac{1}{3}$.

We assume now as in II.4. that $p = p_Z \otimes p_c$. In order to state theorem 3 we changed $U(\sigma) = Y(\sigma) + y(\sigma)$ where $y(\sigma)$ is distributed with respect to p_c.

Define (cf. II.4.)

$$K = F\gamma(\sigma) = \varkappa$$

$$\tilde{K} = E\varkappa(\sigma) * \gamma(\sigma)^+$$

$$\tilde{\tilde{K}} = E\gamma(\sigma) * \gamma(\sigma)^+ = \tilde{\tilde{\varkappa}}.$$

Then

$$\tilde{\varkappa} = p_c * \tilde{K}$$

and

(3) $\partial^2 R = e^{-c\tau}\left(c\tilde{\tilde{K}} + cK * \frac{\delta}{\delta' + c\delta - cH} * c\tilde{K} + c\tilde{K}^+ * \left[\frac{\delta}{\delta' + c\delta - H}\right]^+ * cK^+\right).$

This is the analogue for theorem 3.

In order to state the analogue of theorem 4 introduce

$$\Gamma\mathfrak{m} = \gamma\mathfrak{m}$$

$$\tilde{\Gamma}\mathfrak{m} = A\mathfrak{m} * \Gamma\mathfrak{m}^+$$

$$\tilde{\tilde{\Gamma}}\mathfrak{m} = \Gamma\mathfrak{m} * \Gamma\mathfrak{m}^+$$

and define Γ, $\tilde{\Gamma}$, $\tilde{\tilde{\Gamma}}$ in the same way as B, $\tilde{B}$, $\tilde{\tilde{B}}$ has been defined in II.4. Then we get for $c\tau < \log 2$

(4) $\partial^2 R = c\tilde{\tilde{\Gamma}} + c\Gamma * \frac{\delta}{\delta' + c\delta - cA} * c\tilde{\Gamma} + c\tilde{\Gamma}^+ * \left(\frac{\delta}{\delta' + c\delta - cA}\right)^+ * c\Gamma^+ .$

Literature

[1] Anderson, P.W. and Talman, J.D.: Pressure broadening of spectral lines at general pressures. Conference on the broadening of spectral lines. University of Pittsburgh, Sept. 15-17, 1955, pp. 29-61.

[2] Griem, H.R.: Plasma Spectroscopy. New York, McGraw Hill, 1964.

[3] Waldenfels, W. von: Zur mathematischen Theorie der Druckverbreitung von Spektrallinien. Z. Wahrscheinlichkeitstheorie verw. Geb. 6, 65-112 (1966).

MOMENTS OF POINT PROCESSES

Klaus Krickeberg

Universität Heidelberg

Introduction

The present article grew out of a series of lectures given by the author in spring 1970 in the joint seminar on probability theory of McMaster University and the Université de Montreal, and at various other Canadian Universities.* The subject of these talks had been the theory of the correlation measure of second order stationary line processes as given by R. Davidson (3) and the author (5). In comparison, the article was expanded to treat also higher moments in a systematic way. To reduce invariance properties of point process and their moment measures to well known properties of measures invariant under certain groups which arise in geometry may be regarded as its main theme. This allows to derive in full generality propositions like theorem 2, corollary, which had been obtained in particular cases by R. Davidson using elementary methods, and theorem 6 which, in the case $k = n = 2$, had been conjectured by the author and then proved by Davidson.

While writing these notes the author learned of the untimely death of Rollo Davidson. During the short time he had worked in this

*The author is greatly indebted to many Canadian colleagues, in particular Prof. M. Behara, for organizing this seminar, and to the National Research Council and various Canadian Universities for financing it. The article was written while the author was a visiting professor at the University of Buenos Aires under its multinational program where he had the benefit of stimulating discussions with Prof. L. A. Santalo.

domain of intriguing problems on the border line of geometry and probability theory, his great imagination contributed a wealth of new insights of which the present article is only one testimony among many.

§ 1. Disintegration of Invariant Measures

Let Y be a locally compact space and $\mathcal{H}$ a locally compact group which acts continuously on Y [2, § 2, n°4] . We denote by ~ the equivalence relation determined by $\mathcal{H}$, that is, $\xi \sim \eta$ if and only if $\eta = H\xi$ for some $H \in \mathcal{H}$. We make the following assumptions: Y and $\mathcal{H}$ have countable bases; there exists a Borel representation of ~ , that is, a locally compact space Γ with a countable base and a map r of Y onto Γ such that a subset Δ of Γ is borelian if and only if $r^{-1}(\Delta)$ is borelian in Y. In particular, each equivalence class $Y_\gamma = r^{-1}\{\gamma\}$ will be a Borel set.

Note that in general Γ cannot be the quotient space Y/~ endowed with the quotient topology because the latter may not be separated. It would be interesting to know whether a Borel representation always exists in the present context.

Next we assume that there is a non-negative bounded Baire function b on Y with the following properties: for every $\gamma \in \Gamma$, the set $Y_\gamma \cap \{b>o\}$ contains a non-empty subset which is open in Y_γ; for every compact subset Δ of Γ ,the set $r^{-1}(\Delta) \cap$ carrier (b) is relatively compact. Again it may be that this is always true.

In the following the term "measure" will always mean "positive Radon measure".

Our final, and crucial, assumption bears on the action of $\mathcal{H}$ in the various equivalence classes: for every γ there is an $\mathcal{H}$-invariant measure τ_γ in Y concentrated on Y_γ, and only one, up to a non-negative factor. We can then normalize τ_γ by requiring that $\tau_\gamma(b)$ be a Baire function of γ, bounded on every compact set, and $\tau_\gamma(b) > 0$ unless $\tau_\gamma = 0$.

We are now in a position to describe $\mathcal{H}$-invarient measures in Y in terms of measures in Γ:

<u>Theorem 1</u>. [5, § 1]. A measure ν in Y is $\mathcal{H}$-invariant if and only if there exists a measure κ in Γ such that the family $(\tau_\gamma)_{\gamma \in \Gamma}$ becomes scalarly κ-integrable, and

(1.1) $$\nu = \int_\Gamma \tau_\gamma \, \kappa(d\gamma).$$

κ is uniquely determined by ν.

Recall that (1.1) amounts to $\nu(f) = \int_\Gamma \tau_\gamma (f) \, \kappa \, (d\gamma)$ for every ν-integrable function f.

The first application of this disintegration of an $\mathcal{H}$-invariant measure ν concerns its invariance under maps not in $\mathcal{H}$:

<u>Theorem 2</u>. [5, § 1]. Let F be a homeomorphism of Y which induces a bijective transformation Φ of Γ such that $F(Y_\gamma) = Y_{\Phi(\gamma)}$ for every γ, or in other words

(1.2) $$r \circ F = \Phi \circ r.$$

Suppose in addition that for every γ the measure $\tau_{\Phi(\gamma)}$ is the image of τ_γ under F. Then an $\mathcal{H}$ invariant measure ν in Y represented in the form (1.1) is invariant under F if and only if the corresponding measure κ in Γ is invariant under Φ.

Note that Φ and Φ^{-1} are necessarily Borel maps. The existence of a transformation Φ which satisfies (1.2) is assured, for example, if $FHF^{-1} \in \mathcal{H}$ for every $H \in \mathcal{H}$.

In the case where each τ_γ is finite we may, of course, use a constant normalization, more precisely

(1.3) $$\tau_\gamma \, (Y) = \beta$$

with some $\beta > 0$ for all γ such that $\tau_\gamma \neq 0$. Then if g denotes a κ-integrable function on Γ, it follows from (1.1) and (1.3) that

(1.4) $$\kappa(g) = \beta^{-1} \nu(g \circ r),$$

in particular

$$\kappa(\Delta) = \beta^{-1} \nu(r^{-1}\Delta)$$

for every Borel subset Δ of Γ.

2. Factorized Invariant Measures

We pass to the particular case

(2.1) $\qquad Y = Q \times T$

where Q and T, too, are locally compact spaces with a countable base. Consider a borelian subset Γ^o of Γ with the following property: $(q,\theta) \sim (q',\theta)$ for all q, $q' \in Q$ such that $r(q,\theta) \in \Gamma^o$. Let T^o be the set of all $\theta \in T$ which satisfy $r(q,\theta) \in \Gamma^o$ for one, and hence for all, $q \in Q$. Clearly

$$r^{-1}(\Gamma^o) = Q \times T^o$$

is a Borel set in Y, and therefore T^o is borelian in T. Moreover, $Q \times T^o$ is invariant under $\mathcal{H}$.

Let $\theta, \theta' \in T^o$. Then we have

(2.2) $\qquad (q,\theta) \sim (q',\theta')$

for one pair q, $q' \in Q$ if and only if this holds for all pairs q, q'. In fact (2.2) implies $(p,\theta) \sim (q,\theta) \sim (q',\theta') \sim (p',\theta')$ for all p, $p' \in Q$. In this way (2.2) defines an equivalence relation in T^o to be denoted again by $\sim$ for the sake of simplicity. Thus $(q,\theta) \sim (q',\theta')$ and $\theta \in T^o$ imply $\theta' \in T^o$ and $\theta \sim \theta'$, and conversely, $\theta, \theta' \in T^o$ and $\theta \sim \theta'$ entail $(q,\theta) \sim (q'\ \theta')$. Hence each equivalence class Y_γ with $\gamma \in \Gamma^o$ induces an equivalence class T_γ in T^o which in turn determines Y_γ by

$$Y_\gamma = Q \times T_\gamma .$$

Next assume that for every $\gamma \in \Gamma^o$ the measure τ_γ can be factorized in the form

$$\tau_\gamma = \rho \otimes \sigma_\gamma$$

where ρ is some fixed measure in Q and σ_γ a measure in T carried by T_γ . Then the representation (1.1) of an $\mathcal{H}$-invariant measure ν in Y for factorized functions

$$(f \otimes g)(q,\theta) = f(q)\ g(\theta)$$

takes the form

$$\nu(f \otimes g) = \rho(f) \int_{\Gamma^o} \sigma_\gamma(g)\ \kappa\ (d\gamma) + \int_{\Gamma - \Gamma^o} \tau_\gamma(f \otimes g)\ \kappa\ (d\gamma)$$

which is valid, for example, for any bounded Baire functions f on Q and g on T with compact carriers.

In particular, if $\nu(Q \times (T - T^o)) = 0$, or equivalently $\kappa\ (\Gamma - \Gamma^o) = 0$, we get

$$\nu(f \otimes g) = \rho(f) \int_{\Gamma^o} \sigma_\gamma(g)\ \kappa\ (d\gamma). \tag{2.3}$$

§ 3. Products of Identical Measures

We start from a locally compact space X with a countable base, and denote by $\mathfrak{K}(X)$ the space of all continuous functions on X with a compact carrier, and by $\mathfrak{B}_o(X)$ the class of all relatively compact Borel subsets of X. Let μ be a measure in X. We employ as before the notation $\mu(f)$ for $f \in \mathfrak{K}(X)$ as well as for more general μ-integrable functions, and $\mu(A)$ for $A \in \mathfrak{B}_o(X)$ as well as for a general μ-measurable A. The measure μ is called diffuse if $\mu\ \{\xi\} = 0$ for every single point ξ of X, and a point measure if it is carried by a finite or countable set, assigning measure 1 to each point of that set.

Let k be a positive integer and $Y = X^k$. We write $\mu^{(k)}$ for the k-th power of μ. Thus $\mu^{(k)}$ is the measure in Y defined by

$$\mu^{(k)}\ (f_1 \otimes \ldots \otimes f_k) = \mu(f_1) \ldots \mu(f_k) \text{ for } f_1, \ldots, f_k \in \mathfrak{K}(X).$$

Therefore, $\mu^{(k)}$ is symmetric, that is, invariant under all permutations of the axes of X^k.

Clearly, if μ is diffuse or a point measure, $\mu^{(k)}$ also is.

Consider a partition $\mathfrak{J} = \{J_1, \ldots, J_m\}$ of the set $\{1, \ldots, k\}$ into m disjoint non-empty subsets $J_1, \ldots, J_m$. We define the "$\mathfrak{J}$-diagonal" $D_{\mathfrak{J}}$ of Y to be the set of all $(\xi_1, \ldots, \xi_k)$ of Y such that, for every $j = 1, \ldots, m$, we have $\xi_i = \xi_{i'}$ for all $i, i' \in J_j$. By the projection $\Pi_{\mathfrak{J}}$ of $D_{\mathfrak{J}}$ onto the space X^m we mean the map $\Pi_{\mathfrak{J}}(\xi_1 \ldots, \xi_k) = (\eta_1 \ldots, \eta_m)$ where $\eta_j = \xi_i$ for all $i \in J_j$. This definition of $\Pi_{\mathfrak{J}}$ is, of course, dependent on the fact that we have arranged the sets $J_1, \ldots, J_m$ in a definite order, whereas $D_{\mathfrak{J}}$ is not, but any other order of the J_j's will only amount to a permutation of the axes of X^m. Therefore we may be excused for the sloppy notation.

Obviously, $\Pi_{\mathfrak{J}}$ is bijective, and a subset C of $D_{\mathfrak{J}}$ is borelian if and only if $\Pi_{\mathfrak{J}} C$ is.

To simplify the notations in the following computation we will treat the case where $j < j'$ and $i \varepsilon J_j$, $i' \varepsilon J_{j'}$ imply $i<i'$; the general case can be reduced to this one by a permutation of the axes of X^k. Let ℓ_j be the number of elements of J_j. Given any sets $B_{ji} \varepsilon \mathcal{B}_0(X)$ for $j = 1, \ldots, m$ and $i = 1, \ldots, \ell_j$ we have

$$D_{\mathfrak{J}} \cap (B_{11} \times \ldots \times B_{1\ell_1} \times \ldots \times B_{m1} \times \ldots \times B_{m\ell_m})$$

(3.1)

$$= D_{\mathfrak{J}} \cap ((B_{11} \cap \ldots \cap B_{1\ell_1})^{\ell_1} \times \ldots \times (B_{m1} \cap \ldots \cap B_{m\ell_m})^{\ell_m}).$$

Hence the part concentrated on $D_{\mathfrak{J}}$ of any measure in Y is completely determined by the values of this measure for sets C of the form

$$C = D_{\mathfrak{J}} \cap (B_1^{\ell_1} \times \ldots \times B_m^{\ell_m}) \tag{3.2}$$

with $B_j \varepsilon \mathcal{B}_0(X)$ which have the projection

$$\Pi_{\mathfrak{J}} C = B_1 \times \ldots \times B_m . \tag{3.3}$$

To compute the part of $\mu^{(k)}$ carried by $D_{\mathfrak{J}}$ we make use of Fubini's theorem. In the first step we get for a set of the type (3.2):

$$\mu^{(k)}(C) = \prod_{j=1}^{m} \mu^{(\ell_j)}(\tilde{B}_j)$$

where $\tilde{B}_j = \{(\xi_1, \ldots, \xi_{\ell_j}): \xi_1 = \ldots = \xi_{\ell_j} \varepsilon B_j\}$. To evaluate further the j-th factor of this product we have to distinguish the cases $\ell_j = 1$ and $\ell_j > 1$. In the first case, this factor is clearly equal to $\mu(B_j)$. In the second case we find, again by Fubini's theorem,

$$\mu^{(\ell_j)}(\tilde{B}_j) = \int_{B_j} \mu\{\xi\}^{\ell_j - 1}\,\mu(d\xi) = \sum_{\xi\varepsilon B_j} \mu\{\xi\}^{\ell_j}$$

where it suffices, of course, to extend the sum over all points $\xi\varepsilon B_j$ such that $\mu\{\xi\} > 0$. Thus

(3.4) $$\mu^{(k)}(C) = \prod_{j:\ell_j=1} \mu(B_j) \prod_{j:\ell_j>1} \Big(\sum_{\xi\varepsilon B_j} \mu\{\xi\}^{\ell_j}\Big).$$

<u>Theorem 3.</u> Suppose that $m<k$. Then μ is diffuse if and only if $\mu^{(k)}(D_{\mathcal{J}}) = 0$. It is a point measure if and only if

(3.5) $$\mu^{(k)}(C) = \mu^{(m)}(\Pi_{\mathcal{J}} C)$$

for every Borel subset C of $D_{\mathcal{J}}$.

Proof. The assumption $m<k$ is equivalent to $\ell_j>1$ for at least one j. Therefore the first assertion follows immediately from (3.4).

Next, suppose that μ is a point measure. Then $\mu\{\xi\} = 1$ for every $\xi\varepsilon X$ such that $\mu\{\xi\} > 0$, hence (3.5) follows from (3.3) and (3.4).

Conversely, suppose that (3.5) is true. The case $\mu = 0$ being trivial we select a set A with $\mu(A) > 0$. Let B be any set in $\mathcal{B}_0(X)$, and define $B_j = A$ if $\ell_j = 1$ and $B_j = B$ if $\ell_j > 1$. Applying (3.5) with the set C given by (3.2) we obtain on account of (3.3) and (3.4):

(3.6) $$\mu(B)^{\ell} = \prod_{j:\ell_j>1} \Big(\sum_{\xi\varepsilon B} \mu\{\xi\}^{\ell_j}\Big)$$

where $\ell = \sum_{j:\,\ell_j>1} 1$. Taking for B a one-point set $\{\eta\}$ with any $\eta\varepsilon X$ we find that $\mu\{\eta\}^{\ell} = \mu\{\eta\}^{\ell'}$ where $\ell' = \sum_{j:\,\ell_j>1} \ell_j > \ell$, hence $\mu\{\eta\} = 1$ if $\mu\{\eta\} > 0$. Therefore by (3.6), if B is any set in $\mathcal{B}_0(X)$, we have $\mu(B)^{\ell} = (\sum_{\xi\varepsilon B} \mu\{\xi\})^{\ell}$ which implies $\mu(B) = \sum_{\xi\varepsilon B} \mu\{\xi\}$. Since every $\mu\{\xi\}$ is equal to 0 or 1 we see that μ is in fact a point measure.

We remark that (3.5) is trivially true for every measure μ if $m = k$.

The intuitive meaning of (3.5) is, of course, that the measure

$\mu^{\textcircled{k}}$ restricted to $D_{\mathfrak{J}}$ is essentially equal to $\mu^{\textcircled{m}}$, more precisely, $\mu^{\textcircled{m}}$ is the image under $\Pi_{\mathfrak{J}}$ of the restriction of $\mu^{\textcircled{k}}$ to $D_{\mathfrak{J}}$. There are various other ways to express this fact, for example

$$\mu^{\textcircled{m}}(B_1 \times \dots \times B_m) = \mu^{\textcircled{k}}(D \cap (B_1^{\ell_1} \times \dots \times B_m^{\ell_m})) \tag{3.7}$$

for all $B_j \in \mathcal{B}_0(X)$, or on the basis of (3.1):

$$\mu^{\textcircled{k}}\left(D_{\mathfrak{J}} \cap \mathop{\times}_{j=1}^{m} \mathop{\times}_{i=i}^{\ell_j} B_{ji}\right) = \mu^{\textcircled{m}}\left(\mathop{\times}_{j=1}^{m} \bigcap_{i=1}^{\ell_j} B_{ji}\right), \tag{3.7'}$$

and in terms of functions instead of sets:

$$\mu^{\textcircled{k}}\left(1_{D_{\mathfrak{J}}} \bigotimes_{j=1}^{m} \bigotimes_{i=1}^{\ell_j} f_{ji}\right) = \mu^{\textcircled{m}}\left(\bigotimes_{j=1}^{m} \prod_{i=1}^{\ell_j} f_{ji}\right) \tag{3.8}$$

for all $f_{ji} \in \mathcal{K}(X)$.

We observe that, the measures $\mu^{\textcircled{k}}$ and $\mu^{\textcircled{m}}$ being symmetric, the formulas (3.5), (3.7), (3.7') and (3.8) hold as well for any other order of the elements of the J_j's.

We denote by $cd(\mathfrak{J})$ the number of sets of the partition $\mathfrak{J}$ of $\{1, \dots, k\}$. Given two partitions $\mathfrak{J}$ and $\mathfrak{J}'$ we write $\mathfrak{J}' \sqsubset \mathfrak{J}$ if $\mathfrak{J}$ is a subpartition of $\mathfrak{J}'$. Clearly $\mathfrak{J}' \sqsubset \mathfrak{J}$ implies $D_{\mathfrak{J}'} \subseteq D_{\mathfrak{J}}$. To avoid trivialities we will henceforth assume that X contains at least two points. Then the converse statement holds: $\mathfrak{J}' \sqsubset \mathfrak{J}$ if $D_{\mathfrak{J}'} \subseteq D_{\mathfrak{J}}$, and in this case $D_{\mathfrak{J}'} = D_{\mathfrak{J}}$ if and only if $\mathfrak{J}' = \mathfrak{J}$, that is, $cd(\mathfrak{J}') = cd(\mathfrak{J})$.

If $cd(\mathfrak{J}) = k$ or $cd(\mathfrak{J}) = 1$, there is only one partition $\mathfrak{J}_{max} = \{\{1\}, \dots, \{k\}\}$ or $\mathfrak{J}_{min} = \{1, \dots, k\}$, respectively. We have $D_{\mathfrak{J}_{max}} = X^k$ and, arranging the sets $\{j\}$ in their natural order, $\Pi_{\mathfrak{J}_{max}} = id$, whereas $D_{\mathfrak{J}_{min}}$ is the usual diagonal and $\Pi_{\mathfrak{J}_{min}}(\xi, \dots, \xi) = \xi$. Clearly $\mathfrak{J}_{min} \sqsubset \mathfrak{J} \sqsubset \mathfrak{J}_{max}$ for any partition $\mathfrak{J}$.

The intersection of any two diagonals $D_{\mathfrak{J}}$ and $D_{\mathfrak{J}'}$ being again a diagonal we can write $D_{\mathfrak{J}} \cap D_{\mathfrak{J}'} = D_{\mathfrak{J}^*}$ where $\mathfrak{J}^* \sqsubset \mathfrak{J}$ and $\mathfrak{J}^* \sqsubset \mathfrak{J}'$, and we have $\mathfrak{J}^* = \mathfrak{J}'$ if and only if $D_{\mathfrak{J}'} \subseteq D_{\mathfrak{J}}$.

Next we define

$$E_{\mathfrak{J}} = D_{\mathfrak{J}} - \bigcup_{\substack{\mathfrak{J}' : \mathfrak{J}' \sqsubset \mathfrak{J} \\ \mathfrak{J}' \neq \mathfrak{J}}} D_{\mathfrak{J}'} . \tag{3.9}$$

In concrete terms, this is the set of all $(\xi_1, \ldots, \xi_k) \varepsilon D_{\mathfrak{J}}$ such that $i \varepsilon J_j$, $i' \varepsilon J_{j'}$ and $j \neq j'$ implies $\xi_i \neq \xi_{i'}$. Then $\bigcup_{\mathfrak{J}} E_{\mathfrak{J}} = Y$ since $D_{\mathfrak{J}\max} = Y$. Moreover, the sets $E_{\mathfrak{J}}$ are mutually disjoint. In fact, suppose that $\mathfrak{J} = \mathfrak{J}'$. Then $D_{\mathfrak{J}} \cap D_{\mathfrak{J}'} = D_{\mathfrak{J}^*}$ where $\mathfrak{J}^* \sqsubset \mathfrak{J}$ and $\mathfrak{J}^* \neq \mathfrak{J}$, hence $E_{\mathfrak{J}} \cap D_{\mathfrak{J}^*} = \emptyset$ by (3.9) and $E_{\mathfrak{J}} \subseteq D_{\mathfrak{J}}$, thus $E_{\mathfrak{J}} \cap D_{\mathfrak{J}'} = \emptyset$ and a fortiori $E_{\mathfrak{J}} \cap E_{\mathfrak{J}'} = \emptyset$.

Note that, by (3.9), $E_{\mathfrak{J}\min} = D_{\mathfrak{J}\min}$.

<u>Theorem 4.</u> Let ν be any measure in $Y = X^k$. Then there exists a unique decomposition of ν of the form

$$\nu = \sum_{\mathfrak{J}} \nu_{\mathfrak{J}} \tag{3.10}$$

where $\nu_{\mathfrak{J}}$ is a measure carried by $D_{\mathfrak{J}}$, and $\nu_{\mathfrak{J}}(D_{\mathfrak{J}'}) = 0$ if $\mathfrak{J}' \sqsubset \mathfrak{J}$ and $\mathfrak{J}' \neq \mathfrak{J}$.

Proof. To prove the uniqueness it suffices to show that the measure $\nu_{\mathfrak{J}}$ in any such decomposition is carried by $E_{\mathfrak{J}}$, since the sets $E_{\mathfrak{J}}$ are mutually disjoint. This, however, follows immediately from (3.9).

To prove the existence, we define

$$\nu_{\mathfrak{J}}(C) = \nu(E_{\mathfrak{J}} \cap C). \tag{3.11}$$

Clearly, $\nu_{\mathfrak{J}}$ is carried by $D_{\mathfrak{J}}$. Moreover, if $\mathfrak{J}' \sqsubset \mathfrak{J}$ and $\mathfrak{J}' \neq \mathfrak{J}$, we have $D_{\mathfrak{J}'} \cap E_{\mathfrak{J}} = \emptyset$ by (3.9), hence $\nu_{\mathfrak{J}}(D_{\mathfrak{J}'}) = 0$ by (3.11).

Combining theorem 4 with theorem 3 we see that in the case of a diffuse measure μ the decomposition (3.10) of $\nu = \mu^{ⓚ}$ is the trivial one, namely, $\mu^{ⓚ}_{\mathfrak{J}\max} = \mu^{ⓚ}$ and $\mu^{ⓚ}_{\mathfrak{J}} = 0$ for all other $\mathfrak{J}$. In the case of a point measure μ, by (3.5) the definition (3.11) takes the form

$$\mu^{ⓚ}_{\mathfrak{J}}(C) = \mu^{ⓜ}(\Pi_{\mathfrak{J}}(E_{\mathfrak{J}} \cap C)) \tag{3.12}$$

where $m = cd(\mathfrak{J})$, in particular

$$\mu^{(k)}_{\mathfrak{Z}_{min}}(C) = \mu(\Pi_{\mathfrak{Z}_{min}}(D_{\mathfrak{Z}_{min}} \cap C)).$$

§ 4. Moments of Random Measures

Let $(\Omega, \mathfrak{F}, \mathbb{P})$ be some probability space. A random measure z in X is a function on $\mathcal{K}(X) \times \Omega$ with the property that $z(f,\omega)$ is a measure as a function of f for fixed ω and a random variable as a function of ω for fixed f. The former will be denoted by $z^I(\omega)$ and the latter by $z(f)$, that is,

$$z^I(\omega) = (f \to z(f,\omega))$$

and

$$z(f) = (\omega \to z(f,\omega)).$$

By the distribution of z we mean the family of the distributions of all random vectors $(z(f_1),\ldots, z(f_n))$ with $f_1,\ldots,f_n \in \mathcal{K}(X)$, or equivalently, of all random vectors $(z(A_1),\ldots,z(A_n))$ with $A_1\ldots,A_n \in \mathcal{B}_o(X)$. Note that we may describe the distribution of $(z(f_1),\ldots, z(f_n))$ by the expectations

$$\mathbb{E}h(z(f_1),\ldots,z(f_n)) = \int_\Omega h(z(f_1,\omega),\ldots,z(f_n,\omega))\ \mathbb{P}(d\omega)$$

where h runs, for example, through all functions in $\mathcal{K}(R^n)$, or all bounded Baire functions on R^n.

The random measure z is said to be diffuse or a point process if $z^I(\omega)$ is diffuse or a point measure, respectively, for $\mathbb{P}$-almost all ω. Intuitively, a point process consists in throwing at random a finite or countable set of points into the space X in such a way that only finitely many of them fall into any given relatively compact set. If ω describes this realization of the random phenomenon in question, these points carry the point measure $z^I(\omega)$.

Let k be a positive integer for each fixed $\omega\in\Omega$ we can form the k-th power $(z^I(\omega))^{ⓚ}$ of the measure $z^I(\omega)$, to be denoted by $f \mapsto z^{ⓚ}(f,\omega)$ which amounts to $z^{ⓚI} = z^{Iⓚ}$. It is clear that $z^{ⓚ}$ represents a random measure in X^k. If z is diffuse or a point measure, $z^{ⓚ}$ has the same property. In the latter case, using the intuitive picture of a point process we may say that $z^{ⓚ}(C)$ is the number of k-tuples of points that fall into the set $C \varepsilon \mathcal{B}_0(X^k)$.

The random measure z is called of k-th order if the expectation $\mathbb{E}z^{ⓚ}(f)$ exists and is finite for every $f \varepsilon \mathcal{K}(X^k)$. This amounts, of course, to $z^{ⓚ}$ being of first order. A necessary and sufficient condition for this to happen is $\mathbb{E}(z(A)^k < \infty$ for every $A \varepsilon \mathcal{B}_0(X)$, and z will then also be of ℓ-th order for every $\ell \leqslant k$.

If z is of k-th order, the functional $f \mapsto \mathbb{E}z^{ⓚ}(f)$ with $f \varepsilon \mathcal{K}(X^k)$ obviously represents a measure in X^k. We will denote this measure by ν_z^k and term it the k-th moment measure of z. Thus $\nu_z^k = \nu^1_{z^{ⓚ}}$ and explicitly for functions in the form of a product:

$$\nu_z^k\ (f_1 \otimes \ldots \otimes f_k) = \mathbb{E}(z(f_1) \ldots z(f_k)).$$

It follows that ν_z^k shares with the measures $z^{ⓚI}(\omega)$ the property of being symmetric.

Suppose that z is of k-th order and $\mathfrak{Z}$ is a partition of $\{1,\ldots,k\}$ into disjoint non-empty sets with $m = cd(\mathfrak{Z}) < k$. Then theorem 3 has the following corallary:

<u>Th.3, corollary 1.</u> Under the preceding assumptions, z is diffuse if and only if $\nu_z^k\ (D_{\mathfrak{Z}}) = 0$.

We can also derive

<u>Th.3, corollary 2.</u> In addition to the preceding assumptions, suppose that for almost all ω and every $A \varepsilon \mathcal{B}(X)$, the number $z(A,\omega)$ is an integer. Then z is a point process if and only if

$$\nu_z^k(C) = \nu_z^m\ (\Pi_{\mathfrak{Z}}\, C) \tag{4.1}$$

for every Borel set $C \subseteq D_{\mathfrak{Z}}$.

In fact, theorem 3 shows that (4.1) is necessary for z to be a point process. On the other hand, by (3.4) we have almost surely $z^{ⓚ}(C,\omega) \geq z^{ⓜ}(\Pi_{\mathfrak{Z}} C,\omega)$ for every C of the form (3.2), and hence for every Borel set $C \subseteq D_{\mathfrak{Z}}$. Therefore (4.1) implies $z^{ⓚ}(C,\omega) = z^{ⓜ}(\Pi_{\mathfrak{Z}} C,\omega)$ for almost all ω where the exceptional set of ω's may now depend on C. However, since X has a countable base, an exceptional set of probability 0 may be chosen to serve for all C, thus $z^{I}(\omega)$ is a point measure for almost all ω by theorem 3, that is, z is a point process.

The assumption that almost all $z(A,\omega)$ are integers cannot be discarded as shown by the example of a one-point set X with $k = 2$, $m = 1$ and z(X) uniformly distributed in the interval $[0, 3/2]$.

By (4.1), the m-th moment measure of a point process is the image under $\Pi_{\mathfrak{Z}}$ of the restriction of its k-th moment measure to $D_{\mathfrak{Z}}$ for any $\mathfrak{Z}$ such that $cd(\mathfrak{Z}) = m$. Hence

<u>Th. 3, corollary 3</u>. The moment measures of order less than k of a point process are completely determined by its moment measure of order k.

Of course, (4.1) can be rewritten in ways analogous to the previous transformations (3.7), (3.7') and (3.8) of (3.5), in particular

$$\nu_z^m (B_1 \times \ldots \times B_m) = \nu_z^k (D_{\mathfrak{Z}} \cap (B_1^{\ell_1} \times \ldots \times B_m^{\ell_m})) \tag{4.2}$$

and

$$\nu_z^k \left(1_{D_{\mathfrak{Z}}} \cdot \bigotimes_{j=1}^{m} \bigotimes_{i=1}^{\ell_j} f_{ji}\right) = \nu_z^m \left(\bigotimes_{j=1}^{m} \prod_{i=1}^{\ell_j} f_{ji}\right). \tag{4.3}$$

Let z again be any random measure in X of k-th order. For fixed ω, we have the decomposition of $z^{ⓚI}(\omega)$ given by (3.10) and (3.11), that is

$$z^{ⓚ}(C) = \sum_{\mathfrak{Z}} z^{ⓚ} (E_{\mathfrak{Z}} \cap C)$$

for every $C \in \mathcal{B}_0(X^k)$. Taking expectations we get

$$\nu_z^k(C) = \sum_{\mathfrak{Z}} \nu_z^k(E_{\mathfrak{Z}} \cap C),$$

and since the measure $C \to \nu_z^k(E_{\mathfrak{Z}} \cap C)$ is carried by $E_{\mathfrak{Z}}$, this is the decomposition of the measure $\nu = \nu_z^k$ defined by theorem 4. In the case of a point process, on account of (4.1), it takes the form

(4.4) $$\nu_z^k(C) = \sum_{\mathfrak{Z}} \nu_z^{cd(\mathfrak{Z})} (\Pi_{\mathfrak{Z}}(E_{\mathfrak{Z}} \cap C)).$$

§ 5. The doubly stochastic Poisson process

Let μ be a fixed measure in X. Then [6, 9] there exists a random measure z in X with the following properties:

I. If $A_1, \ldots, A_n$ are mutually disjoint sets in $\mathcal{B}_0(X)$, the random variables $z(A_1), \ldots, z(A_n)$ are independent.

II. For every $A \in \mathcal{B}_0(X)$ the random variable $z(A)$ has a Poisson distribution with parameter $\mu(A)$, that is,

(5.1) $$\mathbb{P}\{z(A) = m\} = \frac{1}{m!} \mu(A)^m \exp(-\mu(A)), \; m = 0,1,2,\ldots$$

Any random measure z with these properties is called a Poisson process with mean number of points μ.

It follows from II that

(5.2) $$\nu_z^1 = \mu$$

On the other hand, the distribution of z is completely determined by μ. Moreover, z is a point process if and only if μ is diffuse. Given any point process z which satisfies I and (5.2), we can derive the condition II; this is just a slight generalization of the classical Poisson limit theorem.

Next, consider an arbitrary random measure u in X. A random measure z in X is called a doubly stochastic Poisson process with mean number of points u if its distribution is the mixture with respect to $\mathbb{P}(d\omega)$ of the distributions of the various Poisson processes corresponding to the various measures $u^I(\omega)$ with $\omega\in\Omega$. More precisely, denoting for fixed ω by z_ω a Poisson process with mean number of points $u^I(\omega)$ and by $\mathbb{E}_{\omega;f_1,\dots,f_n}$ the expectation with respect to the distribution of the random vector $(z_\omega(f_1),\dots, z_\omega(f_n))$, we should have

$$(5.3) \qquad \mathbb{E}h(z(f_1),\dots, z(f_n)) = \int_\Omega \mathbb{E}_{\omega; f_1,\dots,f_n}(h)\ \mathbb{P}(d\omega)$$

for every $h\in\mathcal{K}(R^n)$, and hence for every bounded Baire function h on R^n. In particular, from (5.1):

$$\mathbb{P}\{z(A) = m\} = \frac{1}{m!}\int_\Omega u(A,\omega)^m \exp(-u(A,\omega))\ \mathbb{P}(d\omega).$$

It can be proved [6] that such a process z always exists and that the distributions of z and of u determine each other completely. Clearly any random variable $z(A)$ with $A\in\mathcal{B}_o(X)$ takes only integral values. Moreover, z is a point process if and only if u is diffuse. Finally, z is of k-th order if and only if u is, and the moment measures up to the k-th order of z and of u can be computed from each other. We have

$$(5.4) \qquad \nu_z^k(f_1\otimes\dots\otimes f_k) = \sum_{m=1}^{k} \sum_{\{J_1,\dots,J_m\}} \nu_u^m\Big(\bigotimes_{j=1}^{m} \prod_{i\in J_j} f_i\Big)$$

where $\{J_1,\dots,J_m\}$ runs through all partitions of $\{1,\dots,k\}$ into m disjoint non-empty subsets. In particular,

$$\nu_z^1(f) = \nu_n^1(f),$$

$$\nu_z^2(f_1\otimes f_2) = \nu_u^2(f_1\otimes f_2) + \nu_u^1(f_1 f_2).$$

Consider a fixed partition $\mathcal{J} = \{J_1,\dots,J_m\}$, and assume that u is diffuse, hence z is a point process. For every $\omega\in\Omega$ the measure in X^k defined by

$$U^*(f_1\otimes\dots\otimes f_k,\omega) = u^{(m)}\Big(\bigotimes_{j=1}^{m} \prod_{i\in J_j} f_i,\omega\Big)$$

is carried by $D_{\mathcal{J}}$. From the first part of theorem 3 it follows that

$u^*(D_{\mathfrak{J}'},\omega) = 0$ if $\mathfrak{J}' \supset \mathfrak{J}$ and $\mathfrak{J}' \neq \mathfrak{J}$. Taking expectations we see that the measure

$$f_1 \otimes \dots \otimes f_k \to \nu_u^m \Big(\bigotimes_{j=1}^{m} \prod_{i \in J_j} f_i \Big)$$

has the same properties. Therefore, by theorem 4, the decomposition (5.4) is identical with the decomposition (3.10) of the measure $\nu = \nu_z^k$. Hence, comparing (5.4) with (4.4), we get

$$\nu_u^m(A) = \nu_z^m(\Pi_{\mathfrak{J}} E_{\mathfrak{J}} \cap A) \tag{5.5}$$

for every $A \in \mathcal{B}_0(X^m)$; observe that $\Pi_{\mathfrak{J}} E_{\mathfrak{J}}$ is the set of all $(\eta_1,\dots,\eta_m)$ with distinct components η_j. This is, in a sense, an inverse formula to (5.4).

A doubly stochastic Poisson process z is called a mixed Poisson process if its mean number of points u has the form $u(A,\omega) = \mu(A)\, a(\omega)$ with a fixed measure μ and a random variable a.

§ 6. Invariance of Random Measures

Suppose a locally compact group $\mathcal{G}$ with a countable base acts continuously in X. A random measure z in X is called strictly stationary if its distribution is invariant under $\mathcal{G}$. By this we mean that for any $f_1,\dots,f_n \in \mathcal{K}(X)$ and any $G \in \mathcal{G}$ the random vectors $(z(f_1),\dots, z(f_n))$ and $(z(f_1 \circ G),\dots,z(f_n \circ G))$ have the same distribution.

Let k be a positive integer. We denote by $\mathcal{H}$ the diagonal group acting in $Y=X^k$ generated by $\mathcal{G}$, that is, the group of all transformations of the form

$$(\xi_1,\dots,\xi_k) \to (G\xi_1,\dots,G\xi_k)$$

with $G \in \mathcal{G}$. We make the assumptions of §1. For simplicity, an $\mathcal{H}$-invariant measure or set in Y will also be termed $\mathcal{G}$-invariant.

Clearly, any diagonal $D_{\mathfrak{Z}}$ of Y is invariant under $\mathcal{G}$, and can thus be written as

(6.1) $$D_{\mathfrak{Z}} = r^{-1}(\Delta_{\mathfrak{Z}})$$

where $\Delta_{\mathfrak{Z}}$ is a borelian subset of the representation space Γ introduced in § 1. It follows that $\mathfrak{Z}' \subset \mathfrak{Z}$ if and only if $\Delta_{\mathfrak{Z}'} \subseteq \Delta_{\mathfrak{Z}}$. Defining

(6.2) $$\Gamma_{\mathfrak{Z}} = \Delta_{\mathfrak{Z}} - \bigcup_{\substack{\mathfrak{Z}' : \mathfrak{Z}' \subset \mathfrak{Z} \\ \mathfrak{Z}' \neq \mathfrak{Z}}} \Delta_{\mathfrak{Z}'}$$

we have, by (6.1) and (3.9),

(6.3) $$E_{\mathfrak{Z}} = r^{-1}(\Gamma_{\mathfrak{Z}}),$$

and the sets $\Gamma_{\mathfrak{Z}}$ form a partition of Γ. Therefore, the disintegration (1.1) of an $\mathcal{H}$-invariant measure ν in Y can be further decomposed into

(6.4) $$\nu = \sum_{\mathfrak{Z}} \int_{\Gamma_{\mathfrak{Z}}} \tau_\gamma \kappa(d\gamma),$$

and by (3.11) and (6.3), this decomposition coincides with the decompositon (3.10) of ν.

Consider a fixed partition $\mathfrak{Z}$ with $cd(\mathfrak{Z}) = m$. The projection $\Pi_{\mathfrak{Z}}$ maps $D_{\mathfrak{Z}}$ in a one-to-one way onto X^m, and by the definition of the diagonal group, the equivalence classes with respect to $\mathcal{H}$ contained in $D_{\mathfrak{Z}}$ correspond via $\Pi_{\mathfrak{Z}}$ to the equivalence classes of X^m with respect to the diagonal group acting in X^m. The $\mathcal{G}$-invariant measure τ_γ with $\gamma \varepsilon \Delta_{\mathfrak{Z}}$ is transformed by $\Pi_{\mathfrak{Z}}$ into a $\mathcal{G}$-invariant measure $\Pi_{\mathfrak{Z}}\tau_\gamma$ in X^m carried by the corresponding equivalence class. Hence, making use of (6.2) we see that we know all the relevant measures τ_γ appearing in decompositions of the type (6.4) for the space X^m if we know them for the space X^k. In fact, the general $\mathcal{G}$-invariant measure ν' on X^m is given by

$$\nu' = \sum_{\mathfrak{Z}' : \mathfrak{Z}' \subset \mathfrak{Z}} \int_{\Gamma_{\mathfrak{Z}'}} \Pi_{\mathfrak{Z}} \tau_\gamma \, \kappa'(d\gamma)$$

where the sets $\Gamma_{\mathfrak{Z}'}$ and the measures τ_γ are the same as in (6.4), and

κ' is some measure in $\bigcup_{\mathfrak{J}':\mathfrak{J}'\leq\mathfrak{J}} \Gamma_{\mathfrak{J}'}$.

In the sequel we will always assume that $\mathcal{G}$ acts transitively on X. This is tantamount to requiring that the ordinary diagonal $D_{\mathfrak{J}\,min}$ consists of exactly one equivalence class, and we will denote the representing point of Γ by δ so that

$$D_{\mathfrak{J}\,min} = E_{\mathfrak{J}\,min} = Y_\delta,\ \Delta_{\mathfrak{J}\,min} = \Gamma_{\mathfrak{J}min} = \{\delta\} \quad .$$

The image τ of τ_δ by $\Pi_{\mathfrak{J}min}$ is then a $\mathcal{G}$-invariant measure in X, and the only one, up to a factor.

A random measure z in X is said to be stationary up to the k-th order if it is of k-th order, and if for every integer $m \leq k$ the moment measure ν_z^m in X^m is $\mathcal{G}$-invariant. We can then represent its moment measure $\nu = \nu_z^k$ in the form (6.4).

Let z be a point process. The moment measures ν_z^m with $m<k$ being given in terms of ν_z^k by (4.2), it suffices to require the $\mathcal{G}$-invariance, that is, the $\mathcal{H}$-invariance, of ν_z^k to ensure k-th order stationarity. Moreover, writing

(6.5) $$\nu_z^k = \sum_{\mathfrak{J}} \int_{\Gamma_{\mathfrak{J}}} \tau_\gamma \ \kappa(d\gamma)$$

it follows from (4.1) and the preceding discussion that the corresponding decomposition of ν_z^m has the form

$$\nu_z^m = \sum_{\mathfrak{J}':\mathfrak{J}'\leq\mathfrak{J}} \int_{\Gamma_{\mathfrak{J}'}} \Pi_{\mathfrak{J}} \ \tau_\gamma \ \kappa(d\gamma)$$

with the same measure κ where $\mathfrak{J}$ may be any partition of $\{1,\ldots,k\}$ such that $cd(\mathfrak{J})=m$. In particular

$$\nu_z^1 = \kappa\,\{\delta\}\,\tau.$$

Next, consider the case of a doubly stochastic Poisson process with mean number of points u. Clearly, z is strictly stationary if and only if u is, and likewise z is stationary up to the k-th order if and only if u is. In the case where u is diffuse, and therefore z a point process, the decompositions (5.4) and (6.5) are identical, hence

$$\nu_u^m = \int_{\Gamma_{\mathcal{J}}} \Pi_{\mathcal{J}} \tau_\gamma \quad \kappa(d\gamma)$$

in accordance with (5.5). In particular,

(6.6) $$\nu_u^k = \int_{\Gamma_{\mathcal{J}\max}} \tau_\gamma \quad \kappa(d\gamma).$$

Finally suppose that X is compact, and therefore Y, too. Then the image μ^{*k} of the k-th power $\mu^{ⓚ}$ of any measure μ in X under the map r is well defined, and

(6.7) $$\mu^{*k}(g) = \mu^{ⓚ}(g \circ r)$$

for every bounded Baire function g on Γ. Let z be any random measure in X. Applying (6.7) to all measures $\mu = z^I(\omega)$ with $\omega \varepsilon \Omega$ and taking expectations we get

(6.8) $$\mathbb{E}(z^{*k}(g)) = \nu_z^k(g \circ r).$$

If the invariant measures τ_γ are normalized by (1.3) and z is stationary up to the k-th order, the formulas (6.8) and (1.4) allow us to express the measure κ of the decomposition (6.5) directly in the form

$$\kappa(g) = \beta^{-1}\, \mathbb{E}\,(z^{*k}(g)).$$

§7. Factorization of Random Measures

We now combine the reasonings of §2 and §6 by assuming that the space X itself is a product

(7.1) $$X = P \times S$$

of two locally compact spaces P and S with countable bases. We also assume that we are given a locally compact group $\mathcal{U}$ of transformations of P with a countable base which admits a non-trivial invariant measure λ in P, and only one, up to a factor. Let $\mathcal{G}$ be a group acting in X as before and $\mathcal{V}$ the group of all transformations V of S with the property that the map $(p,\phi) \to (p,V\phi)$, where $p \varepsilon P$ and $\phi \varepsilon S$, belongs to $\mathcal{G}$. We make the final assumption that any $G \varepsilon \mathcal{G}$ has the form

$$(7.2) \qquad G(p,\phi) = (U_\phi\, p,\ V\phi)$$

where $V \varepsilon \mathcal{V}$ and $U_\phi \varepsilon \mathcal{U}$ for every $\phi \varepsilon S$.

Consider a random measure u in X which can be factored as

$$(7.3) \qquad u = \lambda \otimes y,$$

that is, $u(f \otimes g) = \lambda(f)\, y(g)$ for all $f \varepsilon \mathcal{K}(P)$ and $g \varepsilon \mathcal{K}(S)$, whery y is a random measure in S. Then, by (7.2) and Fubini's theorem, u is strictly stationary under $\mathcal{G}$ if and only if y is strictly stationary under $\mathcal{V}$.

Let k be a positive integer. By (7.1), the space $Y = X^k$ has, upon a permutation of the factors, the form (2.1) with $Q = P^k$ and $T = S^k$. It follows from (7.3) that

$$(7.4) \qquad \nu_u^k = \lambda^{ⓚ} \otimes \nu_y^k$$

if u or, equivalently, y is of k-th order. Hence u is $\mathcal{G}$-stationary up to the k-th order if and only if y is $\mathcal{V}$-stationary up to the same order.

The trivial implication (7.3) $\Rightarrow$ (7.4) admits the following converse:

<u>Theorem 5</u> [4]. Let $k > 1$ and u be a k-th order random measure in X whose k-th moment measure can be written as

$$(7.5) \qquad \nu_n^k = \lambda^{ⓚ} \otimes \nu'$$

with some measure ν' in T. Then u has the form (7.3).

Proof. Let Ω_o be the set of all $\omega \varepsilon \Omega$ such that the measure $u^I(\omega)$ vanishes identically. Then $\Omega_o \varepsilon \mathcal{F}$ and

$$(7.6) \qquad u(f \otimes g,\omega) = u((f \circ U) \otimes g,\omega)$$

for all $f \varepsilon \mathcal{K}(P)$, $g \varepsilon \mathcal{K}(S)$, $U \varepsilon \mathcal{U}$ and $\omega \varepsilon \Omega_o$. Moreover, since P and S have countable bases, there are sequences $f_n \varepsilon \mathcal{K}^+(P)$ and $g_n \varepsilon \mathcal{K}^+(S)$

such that

$$(7.7) \qquad \Omega - \Omega_o = \bigcup_n \{\omega : u(f_n \otimes g_n, \omega) > 0\} .$$

Let $f \in \mathcal{K}(P)$, $g \in \mathcal{K}(S)$, $U \in \mathcal{U}$ and $f' = f \circ U$. From (7.5) and $\lambda(f) = \lambda(f')$ it follows that

$$\mathbb{E}(u(f \otimes g)\ u\ (f' \otimes g)\ u\ (f_n \otimes g_n)^{k-2}$$

$$= \mathbb{E}(u(f \otimes g)^2 \ \ u(f_n \otimes g_n)^{k-2}) = \mathbb{E}(u(f' \otimes g)^2 \ \ u(f_n \otimes g_n)^{k-2}),$$

hence

$$\mathbb{E}((u(f \otimes g) - u(f' \otimes g))^2 \ \ u(f_n \otimes g_n)^{k-2}) = 0$$

and therefore $u(f \otimes g) = u(f' \otimes g)$ almost surely on the set $\{\omega : u(f_n \otimes g_n, \omega) > o\}$. On account of (7.6) and (7.7) this implies $u(f \otimes g) = u(f' \otimes g)$ almost surely. Making use of the fact that P, S and $\mathcal{U}$ have countable bases we find that, for almost all ω, we have $u(f \otimes g, \omega) = u((f \circ U) \otimes g, \omega)$ for all $f \in \mathcal{K}(P)$, $g \in \mathcal{K}(S)$ and $U \in \mathcal{U}$. By the definition of λ, this entails $u(f \otimes g, \omega) = \lambda(f)\ \ y(g,\omega)$, and it follows now immediately that y is a random measure in S.

Let us look at the case where Y, the diagonal group $\mathcal{H}$ and a certain subset Γ^o of Γ satisfy the assumption of § 2. Clearly, if σ_γ is a $\mathcal{V}$-invariant measure in T carried by T_γ, the measure

$$(7.8) \qquad \tau_\gamma = \lambda^{ⓚ} \otimes \sigma_\gamma ,$$

for $\gamma \in \Gamma^o$, is $\mathcal{G}$-invariant and carried by Y_γ, hence (2.3) holds with $\rho = \lambda^{ⓚ}$ for every $\mathcal{G}$-invariant measure ν carried by $Q \times T^o$. In this context, we have

<u>Th. 5, corollary 1</u>. Let $k>1$, and suppose that Y, $\mathcal{H}$ and Γ^o satisfy the assumptions of § 2, and that τ_γ has the form (7.8) for every $\gamma \in \Gamma^o$. Then every k-th order stationary random measure u with the property $\nu_u^k(Q \times (T - T^o)) = 0$ admits a factorization.

8. Hyperplane Processes

In the framework of the preceding chapter we turn to the particular case where $P=R$ is the real line and $S=S_{n-1}$ is the unit hypersphere in the euclidean space R^n, thus $X=R\times S_{n-1}$ is a hyper-cylinder. The points $\xi = (P,\phi)$ of X correspond in a one-to-one fashion to the oriented hyperplanes in R^n, the hyperplane represented by ξ being the set

(8.1) $\{\alpha:\ \alpha\varepsilon R^n,\ \langle\phi,\alpha\rangle+p=0\}$

where $\langle\phi,\alpha\rangle = \sum_{i=1}^{n} \phi_i\,\alpha_i$, and the orientation is determined by the direction of ϕ

As usual we denote by $\mathcal{O}^+$ the group of all rotations of R^n, that is, all orthonormal transformations with determinant 1. Let $\mathcal{G}$ be the group of all transformations of X induced on the hyperplanes by the euclidean motions of R^n. By (8.1) the euclidean motion $\alpha \to V\alpha + \varepsilon$ with $\varepsilon\,\varepsilon\, R^n$ and $V\varepsilon\mathcal{O}^+$ generates in X the map

(8.2) $(p,\phi) \to (p - \langle V\phi,\varepsilon\rangle,\ V\phi)$.

Hence $\mathcal{G}$ is the group of all maps $(p,\phi) \to (p + t(\phi),\ V\phi)$ where $V\varepsilon\mathcal{O}^+$ and t is a linear functional on R^n.

Let $\mathcal{U}$ be the group of all translations of R, and $\mathcal{V}$ the restriction of $\mathcal{O}^+$ to S_{n-1}. By (8.2), $\mathcal{V}$ consists of all transformations V such that $(p,\phi) \to (p,\ V\phi)$ belongs to $\mathcal{G}$, and every $G\ \varepsilon\,\mathcal{G}$ has the form (7.2) where $V\ \varepsilon\,\mathcal{V}$ and $U_\phi\varepsilon\,\mathcal{U}$ for every $\phi\varepsilon S_{n-1}$.

It is well known [8] that there exists an invariant measure τ on X, and only one, up to a factor, namely $\tau=\lambda\otimes\sigma$ where λ is the one dimensional Lebesgue measure in R and σ the surface measure on S_{n-1}.

Next we are going to study the equivalence classes in a product space $Y=X^k$ with $k\leq n$ under the diagonal group defined by $\mathcal{G}$, and the invariant measures carried by them. As noted in § 6 it would suffice to consider only the case k=n; however, in view of the applications it appears to be more practical to take up directly the general case.

Rearranging the factors of X^k as we did in § 7 we have

$$Y = Q \times T, \quad Q = R^k, \quad T = (S_{n-1})^k.$$

Consider a point $\eta^o = (p_1^o, \ldots, p_k^o; \phi_1^o, \ldots, \phi_k^o)$ of Y. The equivalence class Y_γ which contains η^o consists of all points η of the form

$$(8.3) \qquad \eta = (p_1, \ldots, p_k; V\phi_1^o, \ldots, V\phi_k^o), \; p_i = p_i^o - \langle V\phi_i^o, \varepsilon\rangle$$

where εR^n and $V \varepsilon \mathcal{V}$. We distinguish two cases:

I. Rank $(\phi_1^o, \ldots, \phi_k^o) = k$. In this case, for any fixed $V \varepsilon \mathcal{V}$, the numbers $p_1, \ldots, p_k$ given by (8.3) take all real values if ε runs through R^n, in particular $(p_1, \ldots, p_k; \phi_1^o, \ldots, \phi_k^o) \sim (q_1, \ldots, q_k; \phi_1^o, \ldots, \phi_k^o)$ for all $p_1, \ldots, p_k, q_1, \ldots, q_k$. Hence, denoting by Γ^o the set of all γ's which define equivalence classes of this type, we have the situation outlined in §2 and §7. The set T^o is the set of all $(\phi_1, \ldots, \phi_k)$ of rank k, and the equivalence relation induced in T^o is the equivalence under the diagonal group given by $\mathcal{V}$, that is, $(\phi_1, \ldots, \phi_k) \sim (\psi_1, \ldots, \psi_k)$ if and only if there is a transformation V in $\mathcal{V}$ such that $\psi_i = V\phi_i$, $i = 1, \ldots, k$.

Let $\tilde{\sigma}$ be the Haar measure in $\mathcal{O}^+$ normalized by $\tilde{\sigma}(\mathcal{O}^+) = \sigma(S_{n-1})$, and σ_γ^o the image of $\tilde{\sigma}$ under the map

$$(8.4) \qquad V \to (V\phi_1^o, \ldots, V\phi_k^o)$$

of $\mathcal{O}^+$ onto $T\gamma$. Denote by $|\det(\phi_1^o, \ldots, \phi_k^o)|$ the k-dimensional volume of the parallelepiped spanned by $\phi_1^o, \ldots, \phi_k^o$. Then

$$(8.5) \qquad \tau_\gamma = |\det(\phi_1^o, \ldots, \phi_k^o)|^{-1} \; \lambda^k \otimes \sigma_\gamma^o$$

is an $\mathcal{H}$-invariant measure in Y concentrated on Y_γ, and the only one up to a factor [8]. In the case k = n, the map (8.4) is bijective and τ_γ is called "Poincare's kinematic measure"; in the case k = 1 we have $\tau_\gamma = \tau$.

II. Rank $(\phi_1^o,\ldots,\phi_k^o) = m<k$. To simplify matters let us assume that $\phi_1^o,\ldots,\phi_m^o$ are linearly independent, hence

$$\phi_i^o = \sum_{j=1}^{m} a_i \phi_j^o, \qquad i = m+1,\ldots,k$$

with $a_{ij} \varepsilon R$. Then, for any fixed $V \varepsilon \mathcal{V}$, the numbers $p_i = p_i^o - \langle V\phi_i^o, \mathfrak{e}\rangle$ with $i = 1,\ldots,m$ take all real values if $\mathfrak{e}$ runs through R^n, whereas

$$p_i = p_i^o + \sum_{j=1}^{m} a_{ij}(p_j - p_j^o), \; i = m+1,\ldots,k.$$

Thus every element η of Y_γ can be uniquely represented by

(8.6) $\qquad \eta \leftrightarrow p_1,\ldots,p_m, \chi$ where $\chi = (V\phi_1^o,\ldots,V\phi_m^o)$.

As before, let σ_γ^o stand for the image of $\tilde{\sigma}$ under the map $V \to (V\phi_1^o,\ldots,V\phi_m^o)$. Then, the measure

$$|\det(\phi_1^o,\ldots,\phi_m^o)|^{-1}\lambda^{(m)} \otimes \sigma_\gamma^o$$

corresponds via (8.6) to an $\mathcal{H}$-invariant measure τ_γ in Y carried by Y_γ, and this is the only one, up to a factor. The normalizing factor $|\det(\phi_1^o,\ldots,\phi_m^o)|^{-1}$ makes τ_γ independent of the choice of m linearly independent vectors out of the entire sequence $\phi_1^o,\ldots,\phi_k^o$.

Given a partition $\mathfrak{Z}$ of $\{1,\ldots,k\}$ such that $cd(\mathfrak{Z}) = m$, we define $\Gamma_{\mathfrak{Z}}^o$ to be the set of all γ's in $\Delta_{\mathfrak{Z}}$ such that rank $(\phi_1^o,\ldots,\phi_k^o) = m$, in particular $\Gamma_{\mathfrak{Z}max}^o = \Gamma^o$ and $\Gamma_{\mathfrak{Z}min}^o = \Delta_{\mathfrak{Z}min}$. It follows, of course, that $\Gamma_{\mathfrak{Z}}^o \subseteq \Gamma_{\mathfrak{Z}}$. After a suitable permutation of the axes, the equivalence class Y_γ with $\gamma \varepsilon \Gamma_{\mathfrak{Z}}^o$ consists of all points

$$(p_1,\ldots,p_m, p_{m+1},\ldots,p_k; V\phi_1^o,\ldots,V\phi_m^o, V\phi_{m+1}^o,\ldots,V\phi_k^o)$$

where $\phi_1^o,\ldots,\phi_m^o$ are fixed linearly independent vectors in S_{n-1}, V runs through $\mathcal{V}$, $(p_1,\ldots,p_m)$ takes all values in R^m, each ϕ_i with $i>m$ is equal to some ϕ_j with $j\leq m$, and $\phi_i = \phi_j$, $j\leq m<i$ implies $p_i=p_j$.

The application $\Pi_{\mathcal{J}}$ will reduce this case to the situation I with m in the place of k.

Next we consider a few transformations F of Y which, in general, do not belong to $\mathcal{H}$.

1. F has the form

(8.7) $(p_1,\dots, p_k; \ \phi,\dots,\phi_k) \to (p_1 + q_1,\dots, p_k + q_k; \phi_1,\dots, \phi_k)$

with fixed $q_i \varepsilon R$. If the q_i's are all equal, we obtain the map of Y induced by a "translation of the hypercylinder X parallel to its generators".

2. F_2 is the map

$$(p_1,\dots, p_k; \phi_1,\dots,\phi_k) \to (-p_1,\dots,-p_k; \phi_1,\dots,\phi_k).$$

3. F_3 is the map of Y induced by a change of the orientation of all hyperplanes, that is,

$$(p_1,\dots,p_k; \phi_1,\dots,\phi_k) \to (-p_1,\dots,-p_k; -\phi_1,\dots,-\phi_k).$$

4. F is a map of Y induced by a "reflexion" in R^n, that is,

$$(p_1,\dots,p_k; \phi_1,\dots,\phi_k) \to (p_1,\dots,p_k; W\phi_1,\dots,W\phi_k)$$

where W is an orthonormal transformation of R^n with determinant -1.

5. F is a permutation of the axes of Y, that is,

$$(p_1,\dots,p_k; \phi_1,\dots,\phi_k) \to (p_{i_1},\dots,p_{i_k}; \phi i_1,\dots,\phi i_k)$$

with some fixed permutation $i_1,\dots, i_k$ of $1,\dots, k$.

Clearly, if n is even, $F_3 = F_2H$ with some $H \varepsilon \mathcal{H}$, whereas if n is odd, every F of type 4 can be written as $F = F_2F_3H$ with some $H\varepsilon \mathcal{H}$.

In all of these examples we have $FHF^{-1} \varepsilon \mathcal{H}$ for every $H \varepsilon \mathcal{H}$, hence there is a transformation Φ of Γ which satisfies (1.2). We write $\Phi = \Phi_\ell$ if $F = F_\ell$, $\ell = 2,3$. It also follows from the definition of the τ_γ's that $\tau_{\Phi(\gamma)}$ is the image of τ_γ under F. Therefore, given an $\mathcal{H}$-invariant measure ν on Y in its representation (1.1), we can apply theorem 2 to it. This involves, of course, a more detailed study of Φ.

The general discussion of Φ is rather tedious; a complete description in the case n=k=2 was given in [5]. The only trivial, and still slightly useful, remarks we can make here are that for n even, we have $\Phi_3 = \Phi_2$, whereas $\Phi = \Phi_2\Phi_3$ for every F of type 4 if n is odd.

For the applications we have in mind it suffices to consider the case $\gamma \epsilon \Gamma^o_{\mathfrak{J}}$ with some $\mathfrak{J}$. Then from the shape of the elements of Y_γ exhibited above we derive immediately that $\phi(\gamma) = \gamma$ if F has the form (8.7) and $q_i = q_j$ for any i and j in the same component of the partition $\mathfrak{J}$, in particular if F is induced by a translation of the hypercylinder X parallel to its generators, of if F is an arbitrary map of type 1 and $\mathfrak{J} = \mathfrak{J}_{max}$. We also have $\phi_2(\gamma) = \gamma$. Therefore, $\phi_3(\gamma) = \gamma$ if n is even, and $\phi(\gamma) = \phi_3(\gamma)$ for any F of type 4 if n is odd. Moreover, if k<n, we find that $\phi(\gamma) = \gamma$ for every F of type 4. Finally, in the case k=2, denoting by ϕ_5 the map ϕ which corresponds to the permutation

$$F_5: \quad (p_1, p_2; \phi_1, \phi_2) \to (p_2, p_1; \phi_2, \phi_1),$$

we have $\phi(\gamma) = \phi_5(\gamma)$ for every F of type 4.

On account of these remarks, theorem 2 has the following corollary:

<u>Th. 2, corollary</u>. Let ν be an $\mathcal{H}$-invariant measure on Y given by its representation (1.1), that is, (6.4), and suppose that κ is carried by $\bigcup_{\mathfrak{J}} \Gamma^o_{\mathfrak{J}}$. Then ν is invariant under any translation of X parallel to its generators, the map F_2, the map F_3 if n is even, and any map of type 4 if k<n. If n is odd, ν is invariant under F_3 if and only if it is invariant under any, or some, F of type 4. If k = 2, ν is invariant under any, or some, F of type 4 if and only if it is invariant under F_5. Finally, if κ is carried by Γ^o, ν is invariant under any F of type 1.

The last assertion follows, of course, directly from the fact that by (8.5) the disintegration (6.4) of an $\mathcal{H}$-invariant measure ν

carried by $R^k \times T^o$ takes the factored form

(8.8) $$\nu = \lambda^{\circledast k} \otimes \int_{\Gamma^o} \sigma_\gamma \quad \kappa(d\gamma)$$

where $\sigma_\gamma = |\det(\phi_1^o,\ldots,\phi_k^o)|^{-1} \sigma_\gamma^o$ if $(\phi_1^o,\ldots,\phi_k^o) \in T_\gamma$; this is the situation described by (2.3).

By a k-th order non-degenerate stationary point process z in X we mean a k-th order stationary point process in X with the property that for almost all ω and every $m \leq k$, any m distinct elements (p_1,ϕ_1) $,\ldots,(p_m,\phi_m)$ of the carrier of $z^I(\omega)$ have linearly independent directions $\phi_1,\ldots,\phi_m$. This is equivalent to ν_z^k being carried by $\bigcup_{\mathfrak{z}} r^{-1}(\Gamma^o_{\mathfrak{z}})$, hence Th. 2, corollary applies in this case to $\nu = \nu_z^k$. In addition, ν_z^k is of course invariant under any map F of type 5.

Next, from §7, in particular Th. 5, corollary 1, we get

Th. 5, corollary 2. If k>1, any k-th order $\mathcal{G}$-stationary random measure u in X which satisfies

(8.9) $$\nu_u^k(R^k \times (T - T^o)) = 0$$

admits a factorization $u = \lambda \otimes y$ where y is a k-th order ϑ-stationary random measure on S_{n-1}, and

(8.10) $$\nu_y^k\,(T - T^o) = 0$$

Conversely, given any random measure y in S_{n-1}, $u = \lambda \otimes y$ is a random measure in X, and u is strictly or k-th order $\mathcal{G}$-stationary if and only if y has the corresponding property with respect to ϑ. If y is diffuse, u also is. Finally, (8.9) and (8.10) are equivalent.

Note that, by Th.3, corollary 1, the condition (8.9) or (8.10) implies that u or y, respectively is diffuse.

Let z be the doubly stochastic Poisson process with a diffuse and k-th order stationary mean number of points u, and consider the decomposition (6.5) of its covariance measure. Then, ν_u^k can be disintegrated according to (6.6), and since $\Gamma^o \subseteq \Gamma_{\mathfrak{z}\max}$ it follows that u satisfies (8.9) if and only if

$$\kappa(\Gamma_{\mathcal{J}max} - \Gamma^o) = 0,$$

that is,

(8.11) $$\nu_z^k(E_{\mathcal{J}max} \cap (R^k \times (T - T^o))) = 0$$

Assuming this to be true, the covariance measure ν_y^k of the random measure y on S_{n-1} obtained from Th.5, corollary 2, is given by

$$\nu_z^k = \int_{\Gamma^o} \sigma_\gamma \ \kappa(d\gamma).$$

By a simple transformation of the coordinates in X we derive from the preceding factorization of u that, given any fixed line L in R^n, the point process of the intersections of the hyperplanes of z with L is a mixed Poisson process whose mean number of points is a random multiple of the Lebesgue measure on L [3].

Clearly, the condition (8.11) is, for any k-th order stationary point process z, necessary for z to be k-th order non-degenerate.

We now return to the case of an arbitrary k-th order non-degenerate stationary point process z where $k \geq 2$. As usual we denote by $\| \ \|_m$ the norm in the space $\mathcal{L}_m(\mathbb{P})$. Given a borelian subset M of S_{n-1} and an integer i we define

$$M^{(i)} = [i, i+1[\times M$$

and

$$x_i(M) = z(M^{(i)}).$$

The moment measures ν_z^m with $m \leq k$ being invariant under any translation of X parallel to its generators by Th. 2, corollary, we have

(8.12) $$\|x_i(M)\|_m = \|x_o(M)\|_m, \quad i = 0, \pm 1, \ldots,$$

and the sequence $(x_i(M))_{i=1,2,\ldots}$ is stationary in $\mathcal{L}_2(\mathbb{P})$, that is, $\mathbb{E}(x_i(M)\ x_j(M)) = \mathbb{E}(x_{j-i}(M)\ x_o(M))$ for all i and j.

We set

(8.13) $$y_\ell(M) = \frac{1}{\ell} \sum_{i=o}^{\ell-1} x_i(M) = \frac{1}{\ell}\, z([0,\ell[\times M).$$

By von Neumann's mean ergodic theorem, the limit

(8.14) $y(M) = \lim_{\ell\to\infty} y_\ell(M)$

exists strongly in $\mathcal{L}_2(\mathbb{P})$, and

(8.15) $y(M) = \lim_{\ell\to\infty} \frac{1}{\ell} \sum_{i=o}^{\ell-1} x_{j_i}(M)$

for every strictly increasing sequence of integers $j_i \geq 0$. Of course $y(M)$ is only determined up to a change in a set of probability 0.

The definition (8.15) of y shows that $0 \leq y(M)$, and that $y(M_1 \cup M_2) = y(M_1) + y(M_2)$ almost surely if M_1 and M_2 are disjoint. Let M_j be a sequence of Borel sets in S_{n-1} such that $M_j \searrow \phi$. Then $x_o(M_j) \searrow 0$ almost surely, hence strongly in $\mathcal{L}_2(\mathbb{P})$. Therefore, by (8.12), (8.13) and (8.14),

$$\lim_{j\to\infty} \| y(M_j) \|_2 = 0,$$

hence $y(M_j) \to 0$ almost surely since $y(M_j)$ decreases. It follows that $y(M)$ can be defined for every M in such a way that y is a random measure in S_{n-1}.

Consider m borelian subsets $M_1,\ldots,M_m$ of S_{n-1} where $m \leq k$. The sequences $(x_i(M_k))_{i=1,2,\ldots}$ being bounded in $\mathcal{L}_m(P)$ on account of (8.12), there is a sequence of positive integers $(j_i)_{i=1,2,\ldots}$ such that

$$\lim_{\ell\to\infty} \frac{1}{\ell} \sum_{i=o}^{\ell-1} x_{j_i}(M_h), \quad h = 1,\ldots,m$$

exist strongly in $\mathcal{L}_m(\mathbb{P})$, see, e.g. [7]. By (8.15), these limits are equal to $y(M_1),\ldots, y(M_m)$, respectively. Therefore, writing

$$N_h^{(\ell)} = \bigcup_{i=o}^{\ell-1} M_h^{(j_i)} = \Big(\bigcup_{i=o}^{\ell-1} [j_i, j_i + 1[\Big) \times M_h, \quad h=1,\ldots,m,$$

we have

$$y(M_h) = \lim_{\ell\to\infty} \frac{1}{\ell} z(N_h^{(\ell)}), h = 1,\ldots,m,$$

strongly in $\mathcal{L}_m(\mathbb{P})$ which entails

(8.16)
$$\nu_y^m (M_1 \times \ldots \times M_m) = \lim_{\ell\to\infty} \frac{1}{\ell^m} \nu_z^m(N_1^{(\ell)} \times \ldots \times N_m^{(\ell)})$$

by the definition of ν_z^m and ν_y^m.

Since z is k-th order stationary and non-degenerate, we can represent ν_z^k in the form

(8.17)
$$\nu_z^k = \sum_{\mathfrak{J}} \int_{\Gamma_{\mathfrak{J}}^o} \tau_\gamma \; \kappa(d\gamma).$$

As noted in §6, this implies

$$\nu_z^m = \sum_{\mathfrak{J}':\mathfrak{J}'\subseteq\mathfrak{J}} \int_{\Gamma^o_{\mathfrak{J}'}} \Pi_{\mathfrak{J}'} \; \tau_\gamma \; \kappa(d\gamma)$$

where $\mathfrak{J}$ may be any partition of $\{1,\ldots,k\}$ such that $cd(\mathfrak{J}) = m$. Taking into account the form of the invariant measure τ_γ for $\gamma\varepsilon\Gamma^o_{\mathfrak{J}'}$ we find that

$$\nu_z^m(N_1^{(\ell)} \times\ldots\times N_m^{(\ell)}) =$$

$$= \ell^m \int_{\Gamma^o_{\mathfrak{J}}} (\Pi_{\mathfrak{J}}\sigma_\gamma)(M_1\times \ldots \times M_m)\kappa(d\gamma) + \sum_{\substack{\mathfrak{J}':\, \mathfrak{J}'\subseteq\mathfrak{J} \\ \mathfrak{J}'\neq\mathfrak{J}}} \ell^{cd(\mathfrak{J}')} s_{\mathfrak{J}'}$$

where the numbers $s_{\mathfrak{J}'}$ do not depend on ℓ and $\Pi_{\mathfrak{J}}$ is defined in $(S_{n-1})^k$ in the same way as in X^k. It now follows from (8.16) that

(8.18)
$$\nu_y^m(M_1 \times\ldots\times M_m) = \int_{\Gamma^o_{\mathfrak{J}}} (\Pi_{\mathfrak{J}}\sigma_\gamma) \; (M_1\times \ldots\times M_m) \; \kappa(d\gamma).$$

Let $u = \lambda \otimes y$, hence by (8.18):

(8.19)
$$\nu_u^m = \lambda^{(m)} \otimes \nu_y^m = \lambda^{(m)} \otimes \int_{\Gamma^o_{\mathfrak{J}}} \Pi_{\mathfrak{J}}\sigma_\gamma \; \kappa(d\gamma) = \int_{\Gamma^o_{\mathfrak{J}}} \Pi_{\mathfrak{J}}\tau_\gamma \; \kappa(d\gamma).$$

Define $\tilde{z}$ to be the doubly stochastic Poisson process with mean number of points u and write

(8.20)
$$\nu_{\tilde{z}}^k = \sum_{\mathfrak{J}} \int_{\mathfrak{J}} \tau_\gamma \; \tilde{\kappa}(d\gamma)$$

with some measure $\tilde{\kappa}$ in Γ. Then, as shown in § 6, we have

(8.21)
$$\nu_u^m = \int_{\Gamma_{\mathfrak{J}}} \Pi_{\mathfrak{J}}\tau_\gamma\tilde{\kappa}(d\gamma)$$

for any $\mathfrak{J}$ with $cd(\mathfrak{J}) = m \leq k$. Comparing (8.19) and (8.21) we see that κ and $\tilde{\kappa}$ coincide within $\Gamma_{\mathfrak{J}}$ for every $\mathfrak{J}$, thus $\kappa=\tilde{\kappa}$. By (8.17) and (8.20), this amounts to $\nu_z^k = \nu_{\tilde{z}}^k$ which in turn implies $\nu_z^m = \nu_{\tilde{z}}^m$ for $m = 1,\ldots,k$. Therefore, we have the following

<u>Theorem 6</u>. Given any k-th order non-degenerate $\mathcal{G}$-stationary point process z in $X=R \times S_{n-1}$ with $k>1$, there is a doubly stochastic Poisson process $\tilde{z}$ such that $\nu_z^m = \nu_{\tilde{z}}^m$, $m = 1,\ldots,k$.

The proof of this theorem as well as Th. 5, corollary 2 show that the mean number of points of $\tilde{z}$ has the form $u = \lambda \otimes y$ with a random measure y on S_{n-1} which satisfies (8.10).

We note that this theorem, with k=2, does not hold in the case of the space R endowed with the translation group [1, p. 184]. It does not hold either for the space S_{n-1} with the rotation group as shown by the following reasoning.

Let X be a compact space, $\mathcal{G}$ a group acting continuously on it, μ a $\mathcal{G}$-invariant measure in X normalized by $\mu(X)=1$ and z the point process defined by a single point distributed in X according to μ. Then z is strictly stationary and of order k for all k, and

$$\mathbb{E}\,(z(M)^2) = \mathbb{E}(z(M))$$

for every borelian set $M \subseteq X$. Hence there can be no doubly stochastic Poisson process $\tilde{z}$ such that $\nu_z^2 = \nu_{\tilde{z}}^2$, because if there were such a process with mean number of points u, we would have

$$\begin{aligned}\mathbb{E}(z(M)^2) &= \mathbb{E}(\tilde{z}(M)^2) = \mathbb{E}(u(M)^2) + \mathbb{E}\,(u(M)) \\ &= \mathbb{E}(u(M)^2) + \mathbb{E}(\tilde{z}(M)) = \mathbb{E}(u(M)^2) + \mathbb{E}(z(M))\end{aligned}$$

for any M, thus $u(M) = 0$ almost surely and therefore $\tilde{z} = 0$.

Literature

1. Bartlett, M. S.: An introduction to stochastic processes. 2nd ed. Cambridge, England, 1966.

2. Bourbaki, N.: Topologie generale. Chap. 3: Groupes topologiques (Actual. scientifiques et industrielles 1143). 3ieme ed. Paris 1960.

3. Davidson, R.: Thesis, Cambridge, England 1967.

4. Davidson, R.: Construction of line processes: second order properties. Izv. Akad. Nauk Armjan. SSR Ser. Fiz.-Mat. Nauk.

5. Krickeberg, K.: Invariance properties of the correlation measure of line processes. Izv. Akad. Nauk Armjan. SSR Ser. Fiz.-Mat. Nauk.

6. Krickeberg, K.: **The Cox Process. To appear.**

7. Riesz, F. - Sz Nagy, B.: Lecons d'amalyse fonctionnelle (3ieme ed.) Gauthier-Villais, Paris, 1955.

8. Santalo, L. A.: Introduction to integral geometry (Actual. scientifiques et industrielles 1198). Paris 1953.

9. Waldenfels, W. V.: Fur mathematischen Theorie der Druckverbreiterung von Spektrallinien. II. Z. Wahrscheinlickeitstheorie verw. Geb. 13, 39-59 (1969).

ADDITIVE AND NON-ADDITIVE ENTROPIES OF FINITE MEASURABLE PARTITIONS*

M. Behara and P. Nath

McMaster University

Abstract. A geometrical representation of entropy leads to a generalization which, in special cases, reduces to the Shannon entropy and may easily be connected to the Rényi entropy. This, in two dimensions, is here called Parabolic entropy (which may further be generalized to higher dimensions). Parabolic entropy happens to be a particular case of Polynomial entropies of finite measurable partitions. Finally, some conditional entropies are defined and their properties have been studied in order to prepare a background for the study of entropy of endomorphisms etc. in ergodic theory and for proving Shannon-Wolfowitz coding theorem.

*This research work was supported by the NRC (Canada) grant No. A2977.

§ 1. Measurable Partitions and σ-fields

Let $(\Omega, \mathcal{S}, \mu)$ be a measure space with $\mu(\Omega) = 1$. We define a finite partition $\mathcal{A} = \{A_i\}_{i\varepsilon I}$ to be a collection of non-empty measurable sets such that

$$\mu(A_i \cap A_j) = 0 \text{ if } i \neq j, \text{ and } \mu(\Omega - \bigcup_{i\varepsilon I} A_i) = 0$$

Two finite partitions $\mathcal{A}$ and $\mathcal{B} = \{B_j\}_{j\varepsilon J}$ are called equivalent: $\mathcal{A} \sim \mathcal{B}$ (mod 0) if $\forall A\varepsilon \mathcal{A}$, $\exists$ a $B\varepsilon \mathcal{B}$ such that A and B differ from each other by a set of measure zero. If every element of $\mathcal{A}$ is the union of elements of $\mathcal{B}$, then $\mathcal{B}$ is called a refinement of $\mathcal{A}$ and we write $\mathcal{A} \subset \mathcal{B}$. Obviously, by definition, every partition is equivalent to itself and also a refinement of itself. The sum of two partitions $\mathcal{A}$ and $\mathcal{B}$, denoted as $\mathcal{A} \vee \mathcal{B}$, is defined as the least common refinement of $\mathcal{A}$ and $\mathcal{B}$, i.e.

$$\mathcal{A} \vee \mathcal{B} = \{A_i \cap B_j\}_{i\varepsilon I,\ j\varepsilon J}$$

and the definition can be easily extended to any finite number of partitions. Obviously, the operator V is both commutative and associative. Also,

$$\mathcal{A} \subset \mathcal{B},\ \mathcal{C} \subset \mathcal{D} \Rightarrow \mathcal{A} \vee \mathcal{C} \subset \mathcal{B} \vee \mathcal{D}$$

If $\forall$ $A\varepsilon \mathcal{A}$, $B\varepsilon \mathcal{B}$, $\mu(A \cap B) = \mu(A).\mu(B)$, then $\mathcal{A}$ and $\mathcal{B}$ are called independent partitions.

By a σ-field (or simply field), we shall mean a collection of measurable sets closed under complimentation and countable unions. For a given partition $\mathcal{A}$, the collection of unions of sets from $\mathcal{A}$ together with the null set $\emptyset$, is clearly a finite field and hence a σ- field also, which may be denoted by $\tilde{\mathcal{A}}$. Obviously $\tilde{\Omega} = \{\Omega, \emptyset\}$ and is called the trivial field corresponding to the trivial partition $\{\Omega\}$. The elements of $\mathcal{A}$ are called atoms of $\tilde{\mathcal{A}}$. Thus there is 1-1 correspondence between the set of all finite partitions of Ω and the finite fields on Ω.

§ 2. The Function $z_\alpha(t)$

Let us define the following real-valued function*

$$(2.1) \quad z_\alpha(t) = \frac{t - t^\alpha}{1 - 2^{1-\alpha}}, \; t \in (0, 1], \; \alpha \in [0,\infty),$$
$$= 1, \quad \alpha = 0, \; t = 0$$
$$= 0, \quad t = 0, \; \alpha \in (0, \infty)$$

Obviously,

$$(2.2) \quad z_1(t) = -t \log_2 t, \; t \in (0,1],$$
$$= 0, \; t = 0$$

The function $z_\alpha(t)$ possesses the following properties:

(a_1) $z_\alpha(t)$ is non-negative for all $t \in [0,1]$, $\alpha \in (0,\infty)$, with $z_\alpha(\frac{1}{2}) = \frac{1}{2}$. For $\alpha = 0$, $z_0(t) \geq 0$, $t \in [0,1]$, with $z_0(0) = 1$, $z_0(1) = 0$, and $z_0(\frac{1}{2}) = \frac{1}{2}$.

(a_2) $z_\alpha(t)$ is a continuous function of α and t.

(a_3) $z_\alpha(t)$ is a concave function, i.e., [5]

$$(2.3) \quad z_\alpha(t_1 + t_2) \leq z_\alpha(t_1) + z_\alpha(t_2), \; t_1 \geq 0, \; t_2 \geq 0,$$
$$t_1 + t_2 \in [0, 1], \; \alpha \in [0, \infty).$$

---

*This is a generalization of the function $f(t) = 2(t-t^2)$ proposed earlier [3]. A geometric motivation of this function is given in §5 of this paper.

(a_4) (i) $z_0(t)$ is strictly monotonically decreasing function of $t \in [0,1]$.

(ii) $z_1(t)$ is strictly monotonically increasing for $t \in [0, \frac{1}{2})$ and strictly monotonically decreasing for $t \in [\frac{1}{2}, 1]$.

(iii) $z_\alpha(t)$, $\alpha > 0$, $\alpha \neq 1$, is strictly monotonically increasing function of $t \in [0, (\frac{1}{\alpha})^{\frac{1}{\alpha-1}})$ and strictly monotonic decreasing function of $t \in [(\frac{1}{\alpha})^{\frac{1}{\alpha-1}}, 1]$.

Obviously, from (ii) and (iii), $z_1(t)$ and $z_2(t)$ are the only two functions both strictly monotonically increasing or decreasing according as $t \in [0, \frac{1}{2})$ or $t \in [\frac{1}{2}, 1]$.

(a_5) (i) $z_1(t)$ satisfies the functional equation

(2.4) $$z_1(xy) = xz_1(y) + yz_1(y), \quad (x, y \in [0,1])$$

(ii) $z_\alpha(t)$, $\alpha \neq 1$, $\alpha \geqslant 0$, satisfies the functional equation

(2.5) $$z_\alpha(xy) = xz_\alpha(y) + yz_\alpha(x) + (2^{1-\alpha}-1)\, z_\alpha(x)\, z_\alpha(y), \quad (x,y \in [0,1])$$

In particular, for $\alpha \neq 1$,

(2.6) $$z_\alpha(xy) \leqslant xz_\alpha(y) + yz_\alpha(x), \quad \alpha > 1,$$

(2.7) $$z_\alpha(xy) \geqslant xz_\alpha(y) + yz_\alpha(x), \quad 0 \leqslant \alpha < 1,$$

equality in (2.6) and (2.7 being true in the following four cases:

$$x = y = 0; \quad x = y = 1; \quad x = 1, y = \tfrac{1}{2}; \quad x = \tfrac{1}{2}, y = 1;$$

(a_6) (i) $z_o(t)$ has its maximum value unity when $t = 0$ and minimum value zero when $t = 1$.

(ii) $z_1(t)$ has maximum value $\frac{1}{2}$ when $t = \frac{1}{2}$ and minimum value zero when $t = 0$ and 1.

(iii) $z_\alpha(t)$, $\alpha > 0$, $\alpha \neq 0, 1$, have maximum at $t = (\frac{1}{\alpha})^{\frac{1}{\alpha - 1}}$ and the maximum value is

(2.8) $$M(\alpha) = \frac{\alpha - 1}{\alpha(1 - 2^{1-\alpha})} \left(\frac{1}{\alpha}\right)^{\frac{1}{\alpha-1}}$$

Obviously, from (ii) and (iii), $z_1(t)$ and $z_2(t)$ both possess maxima at $t = \frac{1}{2}$ and have the same maximum value which is also 1/2. For all other positive values of α, no doubt that also $z_\alpha(\frac{1}{2}) = \frac{1}{2}$, but the maxima are no longer attained at $t = \frac{1}{2}$. In fact, $M(\alpha) = t =$

$(\frac{1}{\alpha})^{\frac{1}{\alpha-1}}$ implies $2^{\alpha-1} = \alpha$ which is true only when $\alpha = 1$ or 2. This justifies as to why only $z_1(t)$ and $z_2(t)$ have the same maximum value which is attained only at $t = \frac{1}{2}$.

(a_7) (i) $z_0(t) = 1 - t$, $t \in [0, 1]$, represents the equation of a straight line passing through the points (1,0) and (0,1).

(ii) $z_1(t) = -t \log_2 t$, $t \in (0,1]$, $z_1(0) = 0$, has been plotted. see p-36 in [2].

(iii) $z_2(t) = 2(t - t^2)$ represents the equation of the parabola passing through the points $(\frac{1}{2}, \frac{1}{2})$ and (1,0), with its axis as the line $t = \frac{1}{2}$, vertex $(\frac{1}{2}, \frac{1}{2})$, focus $(\frac{1}{2}, \frac{3}{8})$, and directrix $z_2(t) = \frac{5}{8}$. Because of symmetry around its axis, obviously the parabola passes through the origin. In fact the family of parabola with foci $(\frac{1}{2}, c)$, $0<c<1/2$, vertex $(\frac{1}{2}, \frac{1}{2})$, and the directrix $y = z_2(t) = 1 - c$, is given by

$$t^2 - t - 3/4 + 2y(1 - 2c) + 2c = 0$$

and it can be seen that the only parabola passing through this origin is the one which has $c = 3/8$. For positive integral values of α, the graphs of $z_\alpha(t)$ can be drawn.

$z_\alpha(t)$ is a polynomial of degree α except when $\alpha = 0$ and 1. In the former case, $z_\alpha(t)$ is a polynomial of degree 1 whereas in the latter case, it is a logarithmic function.

(a_8) (i) $z_2(t) = z_2(1-t)$, $t \in [0, 1]$.

(ii) $z_1(t) = z_1(1-t)$, only when $t = 0, \frac{1}{2}, 1$.

(iii) $z_\alpha(t) \neq z_\alpha(1-t)$, $\alpha \neq 1, 2$, except when $t = 0, 1$.

In other words, only $z_2(t)$ is symmetric around $t = \frac{1}{2}$. However, if we define

(2.9) $$\phi_\alpha(t) = z_\alpha(t) + z_\alpha(1-t),\ t \in [0, 1],\ \alpha \geqslant 0,$$

then it is clear that $\phi_\alpha(t) = \phi_\alpha(1-t)$, $t \in [0, 1]$. Obviously, for $\alpha \neq 0$, $\phi_\alpha(0) = \phi_\alpha(1) = 0$, $\phi_\alpha(\frac{1}{2}) = 1$.

(iv) $\phi_0(t)$ is a polynomial of degree zero. In fact $\phi_0(t) = 1$, $t \in [0, 1]$.

(v) $\phi_1(t)$ is logarithmic.

(vi) $\phi_2(t)$ and $\phi_3(t)$ are polynomials of degree 2.

In fact, for positive integral values of $\alpha \geqslant 3$, $\phi_\alpha(t)$ is a polynomial of degree $\alpha - 1$ or α according as α is even or odd integer. Though $z_2(t) \neq z_3(t)$, $t \in (0,1)$, still

(2.10) $$\phi_2(t) = \phi_3(t),\ t \in [0,1].$$

But

$$\phi_\alpha(t) \neq \phi_{\alpha-1}(t),\ \alpha = 4, 5, 6, \ldots$$

Note that (by 2.10), $\phi_3(t)$ is also a parabola with vertex $(\frac{1}{2}, 1)$, focus $(\frac{1}{2}, \frac{15}{16})$, axis $t = \frac{1}{2}$, directrix $\phi_2(t) = \frac{17}{16}$, and the latus rectum 1/4. But $z_3(t)$ is not a parabola.

(a_9) (i) $z_2(t) = 2t\, z_0(t)$

(ii) For $\alpha > 2$,

$$z_\alpha(t) = \left(\frac{1-2^{2-\alpha}}{1-2^{1-\alpha}}\right) z_{\alpha-1}(t) + (1-2^{1-\alpha})^{-1}\, t^{\alpha-1}\, z_0(t)$$

(a_{10}) Since

$$\log_2 t^{\alpha-1} \leq (t^{\alpha-1}-1)\log_2 e, \ \alpha \geq 0 \tag{2.12}$$

equality being true when $\alpha = 1$ or $t = 1$, it follows that

$$z_1(t) \lesseqgtr (\log_2 e)\left(\frac{1-2^{1-\alpha}}{\alpha-1}\right) z_\alpha(t), \ \alpha \lesseqgtr 1. \tag{2.13}$$

Consequently, for $\alpha \neq 0$,

$$\phi_1(t) \lesseqgtr (\log_2 e)\left(\frac{1-2^{1-\alpha}}{\alpha-1}\right)\phi_\alpha(t), \ \alpha \lesseqgtr 1, \tag{2.14}$$

Note that when $\alpha = 0$, $\phi_1(1) = 0$, $\phi_0(1) = 1$. Thus $\phi_1(1) < \phi_0(1)$.

§ 3. Characterization of the function $z_\alpha(t)$

We prove the following theorem to characterize $z_\alpha(t)$.

Theorem 3.1. Let there be two sets $\{a_i\}$, $i = 1,2,\ldots, m$ and $\{b_j\}$, $j = 1, 2, \ldots n$, of non-negative real numbers such that $\sum_{i=1}^{m} a_i = \sum_{j=1}^{n} b_j = 1$. The function $z_\alpha(t)$ satisfies the following functional equation

$$\sum_{i=1}^{m}\sum_{g=1}^{n} z_\alpha(a_i b_j) = \sum_{i=1}^{n} z_\alpha(a_i) + \sum_{j=1}^{m} z_\alpha(b_j) \tag{3.1}$$

$$+ c \sum_{i=1}^{n} z_\alpha(a_i) \cdot \sum_{j=1}^{m} z_\alpha(b_j), \; c = 2^{1-\alpha}-1, \; \alpha \neq 1, \; \alpha \geqslant 0.$$

Conversely, the only continuous solutions of (3.1) satisfying the condition $z_\alpha(\frac{1}{2}) = \frac{1}{2}$ are given by

$$(2.1) \qquad z_\alpha(t) = \frac{t-t^\alpha}{1-2^{1-\alpha}}, \; 0 \leqslant t \leqslant 1, \; \alpha \neq 1, \; \alpha \geqslant 0.$$

Proof: We shall assume that $\alpha \neq 1$ because if $\alpha = 1$, then the last term on the right side in (3.1) vanishes and the resulting functional equation has been solved by Mcleod and Chaundy [6]. Let us choose integers u, r, s, v such that $1 \leqslant r \leqslant u$, $1 \leqslant s \leqslant v$ and write $m = u - r + 1$, $n = v - s + 1$. If we choose

$$a_i = \frac{1}{u}, \; i = 1, 2, 3, \ldots, u - r, \; a_{u-r+1} = \frac{r}{u},$$

$$b_j = \frac{1}{v}, \; j = 1, 2, 3, \ldots, v - s, \; b_{v-s+1} = \frac{s}{v},$$

then it is clear that $\sum_{i=1}^{m} a_i = \sum_{j=1}^{n} b_j = 1$. Moreover a_i and b_j, $i = 1$ to m, $j = 1$ to n, are strictly positive real numbers. Substitution in (3.1) gives

$$(3.2) \begin{cases} (u-r)(v-s)\, z_\alpha(\frac{1}{uv}) + (u-r)\, z_\alpha(\frac{s}{uv}) + (v-s)\, z_\alpha(\frac{r}{uv}) + z_\alpha(\frac{rs}{uv}) \\ = \left[(u-r)\, z_\alpha(\frac{1}{u}) + z_\alpha(\frac{r}{u})\right] + \left[(v-s)\, z_\alpha(\frac{1}{v}) + z_\alpha(\frac{s}{v})\right] \\ + c\left[(u-r)\, z_\alpha(\frac{1}{u}) + z_\alpha(\frac{r}{u})\right] \left[(v-s)\, z_\alpha(\frac{1}{v}) + z_\alpha(\frac{s}{v})\right] \end{cases}$$

Choice of $r = s = 1$ reduces (3.2) to

$$(3.3) \qquad uv\, z_\alpha(\frac{1}{uv}) = u\, z_\alpha(\frac{1}{u}) + v\, z_\alpha(\frac{1}{v}) + c(u\, z_\alpha(\frac{1}{u}))\,(v\, z_\alpha(\frac{1}{v})).$$

Let

$$(3.4) \qquad \psi_\alpha(t) = t\, z_\alpha(\frac{1}{t}), \; t \in (0, \infty]$$

Clearly (3.3) reduces to the following equation

(3.5) $\psi_\alpha(uv) = \psi_\alpha(u) + \psi_\alpha(v) + c\,\psi_\alpha(u).\,\psi_\alpha(v)$, $u \geq 1$, $v \geq 1$, u and v being integers.

Now, let us choose s = 1. Then (3.2) together with (3.3) gives

(3.6) $\psi_\alpha(\frac{uv}{r}) = \psi_\alpha(\frac{u}{r}) + \psi_\alpha(v) + c\,\psi_\alpha(\frac{u}{r}).\ \psi_\alpha(v)$

Similarly, it can be proved that

(3.7) $\psi_\alpha(\frac{uv}{s}) = \psi_\alpha(\frac{u}{r}) + \psi_\alpha(\frac{v}{s}) + c\,\psi_\alpha(u)\ \psi_\alpha(\frac{v}{s})$

Finally, from (3.2), (3.4), (3.5), (3.6) and (3.7), it can be deduced that

(3.8) $\psi_\alpha(\frac{uv}{rs}) = \psi_\alpha(\frac{u}{r}) + \psi_\alpha(\frac{v}{s}) + c\,\psi_\alpha(\frac{u}{r}).\ \psi_\alpha(\frac{v}{s})$.

Since $m \geq 1$, $n \geq 1$, it is clear that $\frac{u}{r}$ and $\frac{v}{s}$ are rational numbers each greater than unity. Thus (3.5) holds for all u and v where u and v are rational numbers greater than unity. Since for any irrational x, we can always find a sequence $\{r_n\}$ of rationals such that $\lim_{n\to\infty} r_n = x$ and then, by the assumption of continuity, it follows that (3.5) is true for all real numbers u and v, $u \geq 1$, $v \geq 1$.

Define

(3.9) $f_\alpha(t) = 1 + c\,\psi_\alpha(t)$, $t \in (0,\infty]$

Obviously, (3.5), now holding for all real numbers u and v, greater than unity, reduces to

(3.10) $f_\alpha(uv) = f_\alpha(u)\ f\ (v)$, $u \geq 1$, $v \geq 1$

whose non-identically vanishing continuous solutions are of the form [1]

(3.11) $f_\alpha(u) = u^\lambda$, $\lambda \in R$, $u \in R$, $u \geq 1$,

so that, from (3.4), (3.9) and (3.11),

(3.12) $z_\alpha(t) = \frac{t^{1-\lambda}\ (1-t^\lambda)}{c}$, $0 \leq t \leq 1$.

Using the fact that $z_\alpha(\frac{1}{2}) = 1/2$, and $c = (2^{1-\alpha}-1)$, it is easily seen that $\lambda = 1 - \alpha$ so that (2.1) follows. Thus, we have proved the second part of the theorem.

To prove the first part, all that we need is to choose $x = a_i$, $y = b_j$ in (2.4) and then sum w.r.t. i and j.

<u>Remark 1</u>: If the condition $z_\alpha(\frac{1}{2}) = \frac{1}{2}$ is not assumed, then it is evident that all the continuous solutions of (3.1) are given by (3.12) and it is possible that, depending upon suitable choices of a and λ, $z_\alpha(t)$ may be even negative. The choice $z_\alpha(\frac{1}{2}) = \frac{1}{2}$, not only enables us to evaluate λ as a function of α, but it rather makes $z_\alpha(t)$ non-negative. It is also essential to assume continuity of $z_\alpha(t)$, otherwise (3.10) will even admit of discontinuous solutions also, and such solutions are of no importance from information-theoretic point of view.

The functional equation (3.10) admits a non-zero constant solution $f(u) = 1$ which corresponds to $\lambda = 0$. But, for $\alpha \neq 1$, (3.12) implies $z_\alpha(t) = 0$, which though satisfies (3.2), is a trivial solution, and is also no longer sensitive to the assumption $z_\alpha(\frac{1}{2}) = 1/2$. When $\lambda \neq 0$, $\alpha \neq 1$, (3.2) admits a non-zero non-constant solutions of the form (3.12). Under such circumstances, $z\alpha(\frac{1}{2}) = \frac{1}{2} \Rightarrow \lambda = 1 - \alpha$. These observations clearly show that the conditions $\lambda = 0$, $\alpha \neq 1$, are not simultaneously compatible with each other. When $\lambda = 0$, $\alpha = 1$, only the non-constant continuous solutions, under the assumption $z_\alpha(\frac{1}{2}) = \frac{1}{2}$, carry the meaning i.e., we have $z_0(t) = 1 - t$ which represents a straight line. If $\lambda \neq 0$ and $\alpha = 1$, it is obviously seen that (3.12) reduces to

$$z_1(t) = \lambda t \log_2 t, \quad 0 \leq t \leq 1,$$

which are solutions (continuous) of (3.2), when $c = 0$. If $z_1(\frac{1}{2}) = 1/2$ is assumed, it reduces to $z_1(t) = -t \log_2 t$.

In addition to (3.11), (3.10) also admits of an identically vanishing solution i.e., $f_\alpha(u) \equiv 0$. But then $z_\alpha(t) = -t/c$. Obviously $z_\alpha(\frac{1}{2}) = \frac{1}{2} \Rightarrow \alpha=1$ which is a contradiction. Thus, this solution is also not desirable. Moreover, $z_1(t) = -\infty$, $t \in (0, 1]$ and $z_1(0)$ is indeterminate.

In fact, the function $z_\alpha(t)$ is defined even for $\alpha < 0$ provided $t \neq 0$. Since, it is a basic necessity that $z_\alpha(t)$ must be defined for all α at $t = 0$, it is not advisable to permit $\alpha < 0$.

§ 4. Entropies of a Measurable Partition

We define the entropy of order α of a finite measurable partition $\mathcal{A}$ as

$$(4.1) \qquad I_\alpha(\mathcal{A}) = \sum_{A \in \mathcal{A}} z_\alpha(\mu(A)) = c(\alpha)(1 - M_\alpha(\mathcal{A})), \quad \alpha \geq 0, \ \alpha \neq 1$$

where

$$(4.2) \qquad c(\alpha) = (1 - 2^{1-\alpha})^{-1}, \quad M_\alpha(\mathcal{A}) = \sum_{A \in \mathcal{A}} \mu^\alpha(A), \quad \alpha \geq 0, \ \alpha \neq 1$$

Obviously,

$$(4.3) \qquad I_1(\mathcal{A}) = \lim_{\alpha \to 1} I_\alpha(\mathcal{A}) = \sum_{A \in \mathcal{A}} z_1(\mu(A)) = -\sum_{A \in \mathcal{A}} \mu(A) \log_2 \mu(A)$$

which is well-known Shannon's entropy [11]. In addition to above entropies, the quantity

$$(4.4) \quad H_\alpha(\mathcal{A}) = (1-\alpha)^{-1} \log_2 M_\alpha(\mathcal{A}), \quad \alpha \geq 0, \ \alpha \neq 1,$$

is called Rényi's entropy of order α. In order to avoid the confusion, we shall call $I_\alpha(\mathcal{A})$ and $H_\alpha(\mathcal{A})$, non-additive and additive entropies of order α of measurable partition $\mathcal{A}$. Clearly

$$(4.5) \qquad I_1(\mathcal{A}) = H_1(\mathcal{A})$$

Theorem 4.1. If $\mathcal{A}$ is a trivial partition, then $I_\alpha(\mathcal{A}) = H_\alpha(\mathcal{A}) = 0$, $\alpha \geq 0$. Conversely, if $I_\alpha(\mathcal{A}) = H_\alpha(\mathcal{A}) = 0$, $\alpha \geq 0$, then $\mathcal{A}$ is a trivial partition.

Proof: (i) Let $\alpha = 1$. If $\mathcal{A} = \{\Omega\}$, then
$I_1(\mathcal{A}) = H_1(\mathcal{A}) = z_1(\mu(\Omega)) = z_1(1) = 0$ by (a_1). Conversely,
$I_1(\mathcal{A}) = H_1(\mathcal{A}) = 0 \Rightarrow -\sum_{A\varepsilon\mathcal{A}} \mu(A)\log_2\mu(A) = 0 \Rightarrow \mu(A)\log_2\mu(A) = 0$
$\forall A \varepsilon \mathcal{A} \Rightarrow \mu(A) = 0$ or $1 \Rightarrow A = \emptyset$ or Ω. Since, by hypothesis, $\mathcal{A}$ consists of non-empty elements, therefore $A = \Omega$ so that $\mathcal{A} = \{\Omega\}$.

(ii) Let $\alpha \neq 1$, $\alpha > 0$. If $\mathcal{A} = \{\Omega\}$, then $M_\alpha(\mathcal{A}) = 1$. Hence $I_\alpha(\mathcal{A}) = H_\alpha(\mathcal{A}) = 0$. Conversely $I_\alpha(\mathcal{A}) = H_\alpha(\mathcal{A}) = 0 \Rightarrow M_\alpha(\mathcal{A}) = 1$
$\Rightarrow \sum_A \mu^\alpha(A) = 1 \Rightarrow \mu(A) = 0$ or $1 \; \forall A \varepsilon \mathcal{A} \Rightarrow A = \emptyset$ or Ω.

By the same argument as in (i), we conclude that $\mathcal{A} = \{\Omega\}$.

(iii) Let $\alpha = 0$. If $\mathcal{A} = \{\Omega\}$, then

$$I_0(\mathcal{A}) = z_0(\mu(\Omega)) = z_0(1) = 0 \text{ by } (a_1)$$

$$H_0(\mathcal{A}) = \log_2 M_0(\mathcal{A}) = \log_2 1 = 0$$

Conversely, if n denotes the number of elements of $\mathcal{A}$, then

$$I_0(\mathcal{A}) = 0 \Rightarrow n - 1 = 0 \Rightarrow n = 1$$

$$H_0(\mathcal{A}) = 0 \Rightarrow \log_2 n = 0 \Rightarrow n = 1$$

Since $\mathcal{A}$ consists of only one element, therefore $\mathcal{A} = \{\Omega\}$.

Lemma 4.1: $H_\alpha(\mathcal{A})$ and $I_\alpha(\mathcal{A})$ are non-negative for $\alpha \geq 0$.

Proof: (i) Let $\alpha = 1$. Both $I_\alpha(\mathcal{A})$ and $H_\alpha(\mathcal{A})$ reduce to

Shannon's entropy whose non-negativity is well-known.

(ii) Let $\alpha = 0$. Then $I_0(\mathcal{A}) = n - 1$ and $H_0(\mathcal{A}) = \log_2 n$. Since $n \geq 1$, $\therefore$ $I_0(\mathcal{A}) \geq 0$, $H_0(\mathcal{A}) \geq 0$.

(iii) Let $\alpha > 0$, $\alpha \neq 1$. The non-negativity of $I_\alpha(\mathcal{A})$ is a consequence of (a_1). Non-negativity of $H_\alpha(\mathcal{A})$ follows from the fact that $M_\alpha(\mathcal{A}) \gtrless 1$, $\alpha \lesseqgtr 1$, $\mathcal{A} \neq \{\Omega\}$.

Lemma 4.2: $\mathcal{A} \sim \mathcal{B}$ (mod0) $\Rightarrow I_\alpha(\mathcal{A}) = I_\alpha(\mathcal{B})$, $H_\alpha(\mathcal{A}) = H_\alpha(\mathcal{B})$, $\alpha \geq 0$. The proof is obvious.

Note that $I_\alpha(\mathcal{A}) = I_\alpha(\mathcal{B}) \not\Rightarrow \mathcal{A} \sim \mathcal{B}$ (mod 0), $H_\alpha(\mathcal{A}) = H_\alpha(\mathcal{B})$ $\not\Rightarrow \mathcal{A} \sim \mathcal{B}$ (mod 0). There are situations where $H_1(\mathcal{A}) = H_1(\mathcal{B}) \not\Rightarrow$ $H_\alpha(\mathcal{A}) = H_\alpha(\mathcal{B})$ and $I_\alpha(\mathcal{A}) = I_\alpha(\mathcal{B})$. For example, consider the partitions $\mathcal{A}$ and $\mathcal{B}$ with probability distributions

$$\mathcal{P}(\mathcal{A}) = \{\tfrac{1}{2}, \tfrac{1}{8}, \tfrac{1}{8}, \tfrac{1}{8}, \tfrac{1}{8}\} \quad , \quad \mathcal{P}(\mathcal{B}) = \{\tfrac{1}{4}, \tfrac{1}{4}, \tfrac{1}{4}, \tfrac{1}{4}\}$$

Clearly $H_1(\mathcal{A}) = H_1(\mathcal{B}) = 2$, but $\mathcal{A} \not\sim \mathcal{B}$ (mod 0). Also $H_\alpha(\mathcal{A}) \neq H_\alpha(\mathcal{B})$, $I_\alpha(\mathcal{A}) \neq I_\alpha(\mathcal{B})$. Meshalkin [8] , by using Shannon's entropy, has proved that the Bernoulli schemes $\{\frac{1}{2}, \frac{1}{8}, \frac{1}{8}, \frac{1}{8}, \frac{1}{8}\}$ and $\{\frac{1}{4}, \frac{1}{4}, \frac{1}{4}, \frac{1}{4}\}$ are insomorphic. Clearly, the above schemes are no longer isomorphic if, instead of Shannon's entropy, we use $H_\alpha(\mathcal{A})$ and $I_\alpha(\mathcal{A})$. We hope to discuss isomorphism problems, by using $H_\alpha(\mathcal{A})$ and $I_\alpha(\mathcal{A})$, in our subsequent work.

Theorem 4.2: $\mathcal{A} \subset \mathcal{B} \Longrightarrow I_\alpha(\mathcal{A}) \leq I_\alpha(\mathcal{B})$, $H_\alpha(\mathcal{A}) \leq H_\alpha(\mathcal{B})$, $\alpha > 0$.

Proof; (i) If $\mathcal{A} = \mathcal{B} = \{\Omega\}$, then $I_\alpha(\mathcal{A}) = I_\alpha(\mathcal{B}) = H_\alpha(\mathcal{A})$ $= H_\alpha(\mathcal{B}) = 0$.

(ii) If $\mathcal{A} = \{\Omega\}$, but $\mathcal{B}$ is not trivial, then $I_\alpha(\mathcal{B}) > 0$, $H_\alpha(\mathcal{B}) > 0$.

(iii) If $\mathcal{A}$ and $\mathcal{B}$ are both not trivial, then if $\mathcal{A} \sim \mathcal{B}$ (mod 0), the answer is given by Lemma 2. Hence, let us assume that $\mathcal{A} \not\sim \mathcal{B}$ (mod 0). Since, by assumption, $\mathcal{B}$ is a refinement of $\mathcal{A}$, therefore each $A \in \mathcal{A}$ is a disjoint union of elements of $\mathcal{B}$. Let $A_i = \bigcup_{j \in J} B_{i,j}$. Clearly, $\mu(A_i) = \sum_j \mu(B_{i,j})$. Hence, by (a_3),

$$z_\alpha(\mu(A_i)) = z_\alpha\left(\sum_j \mu(B_{i,j})\right) \leq \sum_j z_\alpha(\mu(B_{i,j}))$$

Summing both sides over i and using (4.1), it follows that $I_\alpha(\mathcal{A}) \leq I_\alpha(\mathcal{B})$. To prove the other part, we notice that

$$\mu^\alpha(A_i) = \left(\sum_j \mu(B_{i,j})\right)^\alpha \lesseqgtr \sum_j \mu^\alpha(B_{i,j}),\ 0 < \alpha \lesseqgtr 1$$

Summation w.r.t. i and use of (4.4) gives the required result. For $\alpha = 1$, the result has been proved in [5]. When $\alpha = 0$, both $H_0(\mathcal{A})$ and $I_0(\mathcal{A})$ are strictly monotonically increasing functions of n and hence the result is obvious.

The above theorem has the interpretation that the entropy of a paritition increases if a partition is sub-partitioned non-trivially. But how much is the increase in entropy? Clearly, the answer to this question will depend upon as to what type of refinement the partition $\mathcal{B}$ is of the partition $\mathcal{A}$. Hence, whenever $\mathcal{A} \subset \mathcal{B}$, let us be interested in non-negative numbers $x(\alpha)$ and $y(\alpha)$ such that

$$x(\alpha) = I_\alpha(\mathcal{B}) - I_\alpha(\mathcal{A}),\ y(\alpha) = H_\alpha(\mathcal{B}) - H_\alpha(\mathcal{A})$$

Since $\mu(A_i) = \sum_j (B_{i,j}) \Rightarrow \sum_j \frac{\mu(B_{i,j})}{\mu(A_i)} = 1$, let us consider measures ν_i such that,

$$\nu_i(A_i) = 1$$

$$\nu_i(A_j) = 0,\ i \neq j$$

Let $\mathcal{B}_i = \{B_{i,j}\}_{j \in J}$, $B_{i,j} \cap B_{i,k} = \emptyset$, $\bigcup_j B_{i,j} = A_i$. Defining

$$\nu_i(B_{i,j}) = \frac{\mu(B_{i,j})}{\mu(A_i)},$$

it is obvious that

$$\mu^{\alpha}(A_i)\ I_{\alpha}(\mathcal{B}_i) = c(\alpha)\left[\mu^{\alpha}(A_i) - \sum_j \mu^{\alpha}(B_{i,j})\right]$$

so that

$$(4.6)\qquad x(\alpha) = I_{\alpha}(\mathcal{B}) - I_{\alpha}(\mathcal{A}) = \sum_{i\varepsilon I} \mu^{\alpha}(A_i)\ I_{\alpha}(\mathcal{B}_i)$$

From (4.1) and (4.4),

$$(4.7)\qquad I_{\alpha}(\mathcal{A}) = c(\alpha)\ \ (1-2^{(1-\alpha)H_{\alpha}(\mathcal{A})})$$

Hence

$$(4.8)\qquad x(\alpha) = c(\alpha) \sum_{i\varepsilon I} \mu^{\alpha}(A_i)\ (1-2^{(1-\alpha)H_{\alpha}(\mathcal{B}_i)})$$

when $\alpha = 1$, $x(1) = \sum_i \mu(A_i)\ I_1(\mathcal{B}_i)$. In the case of Rényi's entropy.

$$(4.9)\qquad y(\alpha) = (1-\alpha)^{-1}\ \log_2(M_{\alpha}(\mathcal{B})/M_{\alpha}(\mathcal{A}))$$

<u>Theorem 4.3</u>: If $\mathcal{A}$ and $\mathcal{B}$ are independent, then

$$(4.10)\qquad H_{\alpha}(\mathcal{A}\vee\mathcal{B}) = H_{\alpha}(\mathcal{A}) + H_{\alpha}(\mathcal{B}),\ \alpha \geq 0$$

$$(4.11)\qquad I_{\alpha}(\mathcal{A}\vee\mathcal{B}) = I_{\alpha}(\mathcal{A}) + I_{\alpha}(\mathcal{B}) - \frac{1}{c(\alpha)}\ I_{\alpha}(\mathcal{A})\ I_{\alpha}(\mathcal{B}),\ \alpha \geq 0$$

Proof: Since $\mathcal{A}$ and $\mathcal{B}$ are independent, $\mu(A\cap B) = \mu(A).\mu(B)$ $\forall\ A\varepsilon\ \mathcal{A}$, $B\varepsilon\ \mathcal{B}$. Hence

$$M_{\alpha}(\mathcal{A}\vee\mathcal{B}) = M_{\alpha}(\mathcal{A}).\ M_{\alpha}(\mathcal{B})$$

Taking logarithms and making necessary manipulations, (4.10) follows obviously. To prove (4.11), we notice that

$$H_{\alpha}(\mathcal{A}) = \log_2\ (1 - \frac{I_{\alpha}(\mathcal{A})}{c(\alpha)})^{\frac{1}{1-\alpha}},\ \alpha \geq 0,\ \alpha \neq 1$$

Hence (4.10) implies

$$\left(1 - \frac{I_\alpha(\mathcal{A} \vee \mathcal{B})}{c(\alpha)}\right) = \left(1 - \frac{I_\alpha(\mathcal{A})}{c(\alpha)}\right)\left(1 - \frac{I_\alpha(\mathcal{B})}{c(\alpha)}\right)$$

which, upon simplification, reduces to (4.11). In fact, (4.11) is a simple consequence of (2.5) and the fact that $\mu(A \cap B) = \mu(A).\mu(B)$, $A \epsilon \mathcal{A}$, $B \epsilon \mathcal{B}$.

<u>Corollary 1</u>: (i) Let $\alpha = 1$. Both (4.10) and (4.11) reduce to

(4.12) $H_1(\mathcal{A} \vee \mathcal{B}) = H_1(\mathcal{A}) + H_1(\mathcal{B})$,

$\mathcal{A}$ and $\mathcal{B}$ being independent.

(ii) Let $\alpha > 1$. Then (4.11) gives

(4.13) $I_\alpha(\mathcal{A} \vee \mathcal{B}) \leqslant I_\alpha(\mathcal{A}) + I_\alpha(\mathcal{B})$,

equality in (4.13) being true if and only if at least one of $\mathcal{A}$ and $\mathcal{B}$ is trivial. Note that $\mathcal{A}$ and $\mathcal{B}$ are independent if one of them is trivial but not conversely. Thus $I_\alpha(\mathcal{A})$ is sub-additive for $\alpha > 1$.

(iii) Let $0 \leqslant \alpha < 1$. Then (4.11) gives

(4.14) $I_\alpha(\mathcal{A} \vee \mathcal{B}) \geqslant I_\alpha(\mathcal{A}) + I_\alpha(\mathcal{B})$,

equality being true if and only if at least one of $\mathcal{A}$ and $\mathcal{B}$ is trivial. Thus $I_\alpha(\mathcal{A})$ is super-additive when $0 \leqslant \alpha < 1$.

<u>Remark 2</u>: when $\alpha = 1$, it is well-known that

(4.15) $I_1(\mathcal{A} \vee \mathcal{B}) \leqslant I_1(\mathcal{A}) + I_1(\mathcal{B})$,

equality being true if and only if $\mathcal{A}$ and $\mathcal{B}$ are independent. Thus, it is not necessary that at least one of $\mathcal{A}$ and $\mathcal{B}$ must be trivial. If none of $\mathcal{A}$ and $\mathcal{B}$ is trivial, strict inequalities are true in (4.13) and (4.14).

<u>Corollary 2</u>: (i) Let $\alpha = 1$ and $\mathcal{A}_i$, $i = 1,2,\ldots, n$, be mutually independent partitions, then it can be easily seen that

(4.16) $I_1\left(\bigvee_{i=1}^{n} \mathcal{A}_i\right) = \sum_{i=1}^{n} I_1(\mathcal{A}_i)$

(ii) Let $\alpha > 1$ and $\mathcal{A}_i$ be mutually independent partitions. Then

(4.17) $$I_\alpha\left(\bigvee_{i=1}^{n} \mathcal{A}_i\right) \leqslant \sum_{i=1}^{n} I_\alpha(\mathcal{A}_i),$$

equality being true if and only if at most one of $\mathcal{A}_i$, i = 1 to n, is non-trivial.

(iii) Let $0 \leqslant \alpha < 1$, and $\mathcal{A}_i$, i = 1 to n, be mutually independent partitions. Then

(4.18) $$I_\alpha\left(\bigvee_{i=1}^{n} \mathcal{A}_i\right) \geqslant \sum_{i=1}^{n} I_\alpha(\mathcal{A}_i),$$

equality true if and only if at most one of $\mathcal{A}_i$ is non-trivial.

<u>Corollary 3</u>: In general, when $\mathcal{A}_i$'s are mutually independent, and $\alpha \neq 1$,

(4.19) $$\begin{aligned} I_\alpha(\mathcal{A}_1 \vee \mathcal{A}_2 \vee \ldots \vee \mathcal{A}_n) &= \sum_{i=1}^{n} I_\alpha(\mathcal{A}_i) + d(\alpha) \sum_{i \neq j} I_\alpha(\mathcal{A}_i) I_\alpha(\mathcal{A}_j) + d^2(\alpha) \sum_{i \neq j \neq k} I_\alpha(\mathcal{A}_i) I_\alpha(\mathcal{A}_j) I_\alpha(\mathcal{A}_k) \\ &+ \ldots + d^{n-2}(\alpha) \sum_{i_1 \neq i_2 \neq \ldots \neq i_{n-1}} I_\alpha(\mathcal{A}_{i_1}) I_\alpha(\mathcal{A}_{i_2}) \ldots I_\alpha(\mathcal{A}_{i_{n-1}}) \\ &+ d^{n-1}(\alpha) \prod_{i=1}^{n} I_\alpha(\mathcal{A}_i) \end{aligned}$$

where $d(\alpha) = -\frac{1}{c(\alpha)}$. Choosing $\mathcal{A} \sim \mathcal{A}_i$ (mod 0), for all i, (4.19) reduces to

$$I_\alpha(\mathcal{A}_1 \vee \mathcal{A}_2 \vee \mathcal{A}_3 \ldots \vee \mathcal{A}_n) = c(\alpha) - c(\alpha)\left(1 - \frac{I_\alpha(\mathcal{A})}{c(\alpha)}\right)^n, \quad \alpha \neq 1$$

$$= c(\alpha)\left(1 - 2^{n(1-\alpha)H_\alpha(\mathcal{A})}\right), \quad \alpha \neq 1$$

which also follows from (4.7).

<u>Theorem 4.4</u>: The entropy of a partition $\mathcal{A}$ is maximum when all its elements are equiprobable.

Proof: Let n be the number of elements of $\mathcal{A}$ and the partition $\mathcal{E}_n$ whose all elements are equiprobable.

(i) If $\alpha = 1$, then it is well-known that $I_1(\mathcal{A}) < I_1(\mathcal{E}_n)$, equality occuring if and only if $\mathcal{A} \sim \mathcal{E}_n$ (mod 0).

(ii) Let $\alpha > 0$, $\alpha \neq 1$. Then

$$I_\alpha(\mathcal{A}) = \sum_{A \in \mathcal{A}} z_\alpha(\mu(A)) \leq n\, z_\alpha\left(\frac{1}{n}\right) = I_\alpha(\mathcal{E}_n),$$

equality holds if and only if $\mu(A) = \frac{1}{n}$ $\forall A \in \mathcal{A}$ i.e., $\mathcal{A} \sim \mathcal{E}_n$ (mod 0). To prove the fact that $H_\alpha(\mathcal{A}) \leq H_\alpha(\mathcal{E}_n)$, all that is needed is to find the condition under which $M_\alpha(\mathcal{A})$ is maximum when $0 < \alpha < 1$ and minimum when $\alpha > 1$. By elementary calculus, it can be shown that in either situation, $\mu(A) = \frac{1}{n}$, $\forall A \in \mathcal{A}$.

Remark 3: It can be easily seen that $H_1(\mathcal{E}_n) = \log_2 n$ and

(4.20) $$I_\alpha(\mathcal{E}_n) = \frac{1-n^{1-\alpha}}{1-2^{1-\alpha}},\quad H_\alpha(\mathcal{E}_n) = \log_2 n,\quad \alpha > 0,\ \alpha \neq 1$$

Clearly, $H_\alpha(\mathcal{E}_n)$ is strictly monotonically increasing function of n and is independent of α. But $I_\alpha(\mathcal{E}_n)$ depends both upon α and n. Whereas $I_\alpha(\mathcal{E}_n)$ is also strictly monotonically increasing function of n, nothing definite can be said regarding its monotonic character with respect to α. Denoting by $\phi^*_\alpha(n) = 1-n^{1-\alpha}$, it is evident that for $n \geq 2$, $\phi^*_\alpha(n)$ is strictly monotonically increasing function of $\alpha \geq 0$. Since $I_\alpha(\mathcal{E}_n) = \frac{\phi^*_\alpha(n)}{\phi^*_\alpha(2)}$ is a ratio of two strictly monotonic increasing functions of α, it is not necessary that it should also exhibit some monotonic character. If $\mathcal{A} = \mathcal{E}_n$ (mod 0), say $\mathcal{A} = \{A_1, A_2\}$ with $\mu(A_1) = \frac{1}{4}$, then, $I_2(\mathcal{A}) = 0.75$, $I_3(\mathcal{A}) = 0.75$, $I_4(\mathcal{A}) = 0.776$, $I_5(\mathcal{A}) = 0.812$

$$I_{\frac{1}{2}}(\mathcal{A}) = 0.883,\quad I_{\frac{1}{3}}(\mathcal{A}) = 0.917,\quad I_{\frac{1}{4}}(\mathcal{A}) = 0.99$$

Thus $I_\alpha(\mathcal{A})$ exhibits different behaviours in $0 < \alpha < 1$ and $1 < \alpha < \infty$.

From the theoretical point of view, we can also discuss $H_\infty(\mathcal{A})$ and $I_\infty(\mathcal{A})$. From (4.20),

$$I_\infty(\mathcal{E}_n) = 1, \ H_\infty(\mathcal{E}_n) = \log_2 n,$$

so that $I_\infty(\mathcal{E}_n)$ does not even depend upon n. Also, for $n \geq 2$, $I_\infty(\mathcal{E}_n) \leq H_\infty(\mathcal{E}_n)$. On the other hand, $H_0(\mathcal{E}_n) = \log_2 n$, $I_0(\mathcal{E}_n) = n-1$, and it is obvious that $H_0(\mathcal{E}_n) \leq (\log_2 e)\, I_0(\mathcal{E}_n)$, equality holds when $n = 1$ i.e., $\mathcal{E}_n = \mathcal{E}_1 = \{\Omega\}$. For all $\alpha \geq 0$, both $H_\alpha(\mathcal{E}_n)$ and $I_\alpha(\mathcal{E}_n)$ satisfy

$$(4.21) \qquad \lim_{n\to\infty} \frac{I_\alpha(\mathcal{E}_{n+1}) - I_\alpha(\mathcal{E}_n)}{n} = \lim_{n\to\infty} \frac{H_\alpha(\mathcal{E}_{n+1}) - H_\alpha(\mathcal{E}_n)}{n} = 0$$

When $\mathcal{A} \sim \mathcal{E}_n$ (mod 0), $H_\alpha(\mathcal{A})$ is monotonically decreasing function of α. In this respect, $H_\alpha(\mathcal{A})$ differs widely from $I_\alpha(\mathcal{A})$ about which no definite statement as regards its monotony with respect to α can be made. Moreover

$$(4.22) \qquad H_\infty(\mathcal{A}) = \log \frac{1}{\max\limits_{A\varepsilon\mathcal{A}} \mu(A)}, \qquad I_\infty(\mathcal{A}) = 1.$$

Thus $H_\infty(\mathcal{A})$ depends only upon that atom of $\tilde{\mathcal{A}}$ which is most probable whereas $I_\infty(\mathcal{A})$ does not even take into account the probabalistic structure of $\mathcal{A}$.

Lemma 4.3: If T is a measure-preserving transformation, then

$$I_\alpha(T^{-1}\mathcal{A}) = I_\alpha(\mathcal{A}), \ H_\alpha(T^{-1}\mathcal{A}) = H_\alpha(\mathcal{A}).$$

It is a simple consequence of the fact that $\mu(T^{-1}A) = \mu(A)$ $\forall A \varepsilon \mathcal{A}$.

Lemma 4.4: For any finite measurable partition $\mathcal{A}$,

$$(4.23) \qquad H_1(\mathcal{A}) \geq H_\alpha(\mathcal{A}) \geq (\log_2 e) \left(\frac{1 - 2^{1-\alpha}}{\alpha - 1}\right) I_\alpha(\mathcal{A}), \quad \alpha > 1$$

$$(4.24) \qquad H_1(\mathcal{A}) \leq H_\alpha(\mathcal{A}) \leq (\log_2 e) \left(\frac{1 - 2^{1-\alpha}}{\alpha - 1}\right) I_\alpha(\mathcal{A}), \quad 0 \leq \alpha < 1$$

Proof: Due to monotonically decreasing character of $H_\alpha(\mathcal{A})$ w.r.t. α, $H_1(\mathcal{A}) \geqslant H_\alpha(\mathcal{A})$, $\alpha > 1$; and $H_1(\mathcal{A}) \leqslant H_\alpha(\mathcal{A})$, $0 \leqslant \alpha < 1$; equality in both cases being true when either $\mathcal{A}$ is trivial or $\mathcal{A} = \mathcal{E}_n$, $n = 2, 3, \ldots$. Also, since

$$\log_2\left(\sum_A \mu^\alpha(A)\right) \leqslant \left(\sum_A \mu^\alpha(A) - 1\right) \log_2 e = (2^{1-\alpha} - 1)(\log_2 e) I_\alpha(\mathcal{A}),$$

it follows that

$$H_\alpha(\mathcal{A}) \geqslant \left(\frac{1 - 2^{1-\alpha}}{\alpha - 1}\right) (\log_2 e)\, I_\alpha(\mathcal{A}), \quad \alpha > 1.$$

$$H_\alpha(\mathcal{A}) \leqslant \left(\frac{1 - 2^{1-\alpha}}{\alpha - 1}\right) (\log_2 e)\, I_\alpha(\mathcal{A}), \quad 0 \leqslant \alpha < 1$$

Note that equality in (4.23) and (4.24) is true only when $\mathcal{A}$ is trivial.

§ 5. Parabolic Entropy

In this section, we shall strictly assume that $\mathcal{A}$ is a partition with two elements i.e., $\mathcal{A} = \{A_1, A_2\}$, $A_1 \neq \emptyset$, $A_2 \neq \emptyset$, $A_1 \cup A_2 = \Omega$, $\mu(A_1) = p$, $0 < p < 1$. If $p = 0$ or 1, clearly $\mathcal{A}$ reduces to trivial partition. Obviously, by (2.9)

$$I_\alpha(\mathcal{A}) = \phi_\alpha(p)$$

In particular, by (2.10), $I_2(\mathcal{A}) = I_3(\mathcal{A})$. In fact, all the properties of $\phi_\alpha(p)$ are the properties of $I_\alpha(\mathcal{A})$. We shall call $I_2(\mathcal{A})$ as parabolic entropy of $\mathcal{A}$. Note that $H_2(\mathcal{A}) \neq H_3(\mathcal{A})$. For the sake of convenience, let us write

(5.1) $$w_1(p) = -p \log_2 p - (1-p) \log_2(1-p), \quad 0 \leqslant p \leqslant 1$$

(5.2) $$w_\alpha(p) = (1-\alpha)^{-1} \log_2\left[p^\alpha + (1-p)^\alpha\right], \quad 0 \leqslant p \leqslant 1,\ \alpha > 0, \alpha \neq 1.$$

We do not discuss $w_0(p)$ because it does not depend upon p. Due to the monotonic decreasing nature of $w_\alpha(p)$ w.r.t. α, the graphs of $w_\alpha(p)$, $\alpha > 1$, meet the graph of $w_1(p)$ only at the points (0, 0), $(\frac{1}{2}, 1)$ and (1,0), otherwise they lie below the graph of $w_1(p)$. When $0 < \alpha < 1$, the graphs of $w_\alpha(p)$, still pass through the above three points but otherwise lie above the graph of $w_1(p)$. All the graphs are symmetric around the axis $p = \frac{1}{2}$. Actual computation shows that

(5.3) $$w_2(p) = -\log y_1(p),\quad w_3(p) = -\frac{1}{2}\log y_2(p)$$

where

(5.4) $$y_1(p) = 2p^2 - 2p + 1,\quad y_2(p) = 3p^2 - 3p + 1$$

Now we give three different view-points to support the idea as to why we should have parabolic entropy.

(a) The equations $y_1(p) = 2p^2 - 2p + 1$ and $y_2(p) = 3p^2 - 3p + 1$ represent equations of parabolas. More precisely, $y_1(p)$ represents the parabola with vertex $(\frac{1}{2}, \frac{1}{2})$, focus $(\frac{1}{2}, \frac{5}{8})$, axis $p = \frac{1}{2}$, length of latus rectum $= \frac{1}{2}$, directrix $y_1(p) = \frac{3}{8}$, and $y_2(p)$ represents the parabola with vertex $(\frac{1}{2}, \frac{1}{4})$, focus $(\frac{1}{2}, \frac{1}{3})$, etc. In other words, $y_1(p)$ and $y_2(p)$ have the same axis but different vertices and different foci. Rényi's entropies $w_2(p)$ and $w_3(p)$ are functions of $y_1(p)$ and $y_2(p)$ respectively as is obvious from (5.1). Since $I_\alpha(\mathcal{A})$ is connected to $H_\alpha(\mathcal{A})$ by (4.7), it is natural to expect $I_2(\mathcal{A})$ either as a parabola or a function of some parabola. Actual computation gives $I_2(\mathcal{A}) = 4(p-p^2)$ which is a parabola.

Let us consider the image of the parabola $y_1(p) = 2p^2 - 2p + 1$ with respect to the line $y(p) = \frac{1}{2}$. Since $y_1(p) - \frac{1}{2} = \frac{1}{2} - z_2(p)$, obviously the required image is given by the parabola $z_2(p) = 2(p-p^2)$ and $I_2(\mathcal{A}) = 2\, z_2(p)$ because $z_2(p) = z_2(1-p)$. Also, since $y_1(0) = y_1(1) = 1$, and the images of the points (0, 1) and (1, 1) with respect to the lines $y_1(p) = \frac{1}{2}$ are the points (0, 0) and (0, 1), the parabola

$z_2(p)$ must pass through (0, 0), $(\frac{1}{2}, \frac{1}{2})$, (1, 0). Since $w_3(p)$ is also a function of parabola $y_2(p)$, therefore, it is expected that $I_3(\mathcal{A})$ must also be a parabola or a function of a parabola. From (2.3), it is obvious that $I_2(\mathcal{A}) = I_3(\mathcal{A})$. This observation is enough to comment that even in the region $1 < \alpha < \infty$, $I_\alpha(\mathcal{A})$ cannot have a strictly monotonic character with respect to α.

The parabola $z_2(p)$ is not the image of $y_2(p)$ because $y_2(p) - \frac{1}{2} = \frac{1}{2} - \frac{3}{2} z_2(p)$. In fact $y_1(p) = y_2(p) \Rightarrow p = 0$ or $p = 1$. Hence the vertex of $y_2(p)$ must be different from that of $y_1(p)$. It is just a coincidence that $I_2(\mathcal{A}) = I_3(\mathcal{A})$, For $2 < \alpha_1 < \alpha_2 < 3$, $I_{\alpha_1}(\mathcal{A}) \neq I_{\alpha_2}(\mathcal{A})$. Note also that $I_{\alpha_1}(\mathcal{A}) \neq I_{\alpha_2}(\mathcal{A})$, $\alpha_1 \neq \alpha_2$, $\alpha_1 \geqslant 3$, $\alpha_2 \geqslant 3$.

From (5.2), it is easily seen that $w_\alpha(p)$ and $w_{\alpha+1}(p)$ are functions of a polynomial of degree α provided α is an integer. It is natural to ask the following question: Does there exist polynomials as such which represent entropies? The answer is in the affirmative. For $\alpha = 2, 3$, we have already given the answer. Now we discuss this view-point below.

(b) The graph of $w_1(p)$ passes through the points (0, 0), $(\frac{1}{2}, 1)$, (1, 0). Let us find the equation of the parabola $y(p) = ap^2 + bp + c$ which passes through the points (0, 0), $(\frac{1}{2}, \frac{1}{2})$, (1, 0). Obviously $c = 0$, $b = 2$, $a = 1$, so that $y(p) = z_2(p)$. We need the point $(\frac{1}{2}, \frac{1}{2})$ rather than $(\frac{1}{2}, 1)$ for the following reason: By theorem (4.4), $I_\alpha(\mathcal{A})$ is maximum when $p = \frac{1}{2}$. Since (2.3) and (4.1) imply $I_\alpha(\mathcal{A})$ to be a concave function of p, therefore the vertex of the parabola must represent a point of maxima. Since max $I_\alpha(\mathcal{A}) = 1$, therefore the vertex must be $(\frac{1}{2}, \frac{1}{2})$. Since $I_\alpha(\mathcal{A}) = \phi_\alpha(p)$, it follows that $I_2(\mathcal{A}) = 4(p - p^2)$.

It can be shown that there exists no cubic curve $y(p) = -ap^3 - bp^2 - cp + d$ which passes through the points, (0, 0), (1, 0),

$(\frac{1}{2}, \frac{1}{2})$ and has a maximum at $p = \frac{1}{2}$. Actual computation shows that $c = -2$, $b = 2$, $d = 0$, $a = 0$ so that $y(p) = z_2(p)$ again. However if we make the above cubic curve pass through the above three points and possess maxima when $p = \frac{1}{\sqrt{3}}$, with maximum value $\frac{8}{9\sqrt{3}}$, it turns out that $y(p) = z_3(p)$.

Due to the lack of knowledge about boundary conditions, it is difficult to find the exact forms of polynomials of degree α, $\alpha > 0$. However if we assume the polynomial of the form

$$y_\alpha(p) = -Ap^\alpha - Bp + c$$

and make $y_\alpha(p)$ satisfy the conditions $y_\alpha(0) = 0$, $y_\alpha(1) = 0$, $y_\alpha(\frac{1}{2}) = \frac{1}{2}$, then it turns out that $y_\alpha(p) = z_\alpha(p)$. The fact as to why $y_1(p)$ is not defined is quite obvious. There exists no single straight line which passes through the points $(0, 0)$, $(1, 0)$, $(\frac{1}{2}, \frac{1}{2})$. For a fixed p, $y_\alpha(p)$ is a continuous function of α. In fact,

$$\frac{\partial}{\partial\alpha} y_\alpha(p) = - A\, p^\alpha \log p$$

exists for all α so that $y_\alpha(p)$ is a differentiable and hence a continuous function of α. Therefore, it is natural to define

$$y_1(p) = \lim_{\alpha \to 1+0} y_\alpha(p) = \lim_{\alpha \to 1-0} y_\alpha(p)$$

and we have $y_1(p) = z_1(p)$.

If $\alpha = 0$, then $w_0(p) = 1$, $p \in [0, 1]$. Hence $w_0(0) \neq 0$ but $w_0(1) = 1$. Hence the only useful conditions are $y_0(1) = 0$ and $y_0(\frac{1}{2}) = \frac{1}{2}$. But $y_0(p) = -Bp + c$. Making necessary calculations, it follows that $y_0(p) = 1-p = z_0(p)$. Accordingly $I_0(\mathcal{A}) = 1$. Now we discuss the third view-point.

(c) It is obvious that

$$\frac{d^2}{dp^2} w_1(p) = - \frac{1}{p(1-p)} = \psi(p),\ 0 \leq p \leq 1,$$

so that Shannon entropy $w_1(p)$ is such that the second order derivative with respect to p depends upon p. Since $\psi(p) \leq 0$, therefore, $w_1(p)$ is a concave function of p. Let us be interested in a concave function I(p) which represents entropy but has a constant second order derivative, say -k, where $k > 0$, and $I(\frac{1}{2}) = 1$. *Simple integration of $\frac{d^2}{dp^2} I(p) = -k$ gives $I(p) = -\frac{k}{2}p^2 + \frac{k}{2}p$ and $I(\frac{1}{2}) = 1 \Rightarrow k = 8$ so that $I(p) = \phi_2(p)$. There are only two values of p at which $\frac{d^2}{dp^2}\phi_2(p) = \frac{d^2}{dp^2} w_1(p)$. If we use $I(\frac{1}{2}) = \frac{1}{2}$, we get $I(p) = z_2(p)$.

If $\mathcal{A}$ is a partition with n elements, then

$$(5.5) \qquad I_2(\mathcal{A}) = 2(1 - \sum_{A\epsilon} \mu^2(A)) \leq 2,$$

showing that $I_2(\mathcal{A})$ is bounded whatsoever n may be and the bound is independent of n. In this respect, it differs considerably from $H_\alpha(\mathcal{A})$ in which case the upper bound $\log_2 n$ depends upon n. In general,

$$I_\alpha(\mathcal{A}) \leq c(\alpha), \quad \alpha > 1.$$

§ 6. Conditional Entropies

Given a set $A \epsilon \tilde{\mathcal{A}}$ ($\tilde{\mathcal{A}}$ finite) and a finite field $\tilde{\mathcal{B}}$, let $P(A/\tilde{\mathcal{B}})$ denote the essentially unique function defined on Ω, measurable $\tilde{\mathcal{B}}$, such that

$$(6.1) \qquad \int_B P(A/\tilde{\mathcal{B}})d\mu = \mu(A \cap B), \ \forall B \epsilon \tilde{\mathcal{B}}.$$

Clearly, if $A \epsilon \tilde{\mathcal{B}}$, then $P(A/\tilde{\mathcal{B}}) = \chi(A)$, the characteristic function of A, whereas if $\tilde{\mathcal{B}} = \tilde{\Omega}$, then $P(A/\tilde{\mathcal{B}}) = P(A/\tilde{\Omega}) = \mu(A)$. Also, we know that

---

* We use the fact that $I(0) = I(1) = 0$. This requirement is natural if I(p) is to represent entropy.

$$(6.2) \qquad P(A/\tilde{\mathcal{B}}) = \sum_{B \in \mathcal{B}} \frac{\mu(A \cap B)}{\mu(B)} \chi(B) \quad , A \in \tilde{\mathcal{A}} .$$

We define the conditional entropy of $\mathcal{A}$ w.r.t. $\mathcal{B}$ as

$$(6.3) \qquad \hat{I}_\alpha(\mathcal{A}/\mathcal{B}) = \sum_{A \in} \int_\Omega z_\alpha (P(A/\tilde{\mathcal{B}}))d\mu.$$

Clearly, if $\alpha = 1$, this definition reduces to the one given by Brown [5].

Lemma 6.1: $\mathcal{B} = \{\Omega\} \Rightarrow \hat{I}_\alpha(\mathcal{A}/\mathcal{B}) = I_\alpha(\mathcal{A}), \alpha \geq 0.$

Proof: $\mathcal{B} = \{\Omega\} \Rightarrow P(A/\tilde{\mathcal{B}}) = \mu(A)$. Hence $z_\alpha(P(A/\tilde{\mathcal{B}})) = z_\alpha(\mu(A))$ and by (6.3), the result is obvious.

Moreover, $\mathcal{A}$ and $\mathcal{B}$ are independent $\Leftrightarrow$ $\hat{I}_\alpha(\mathcal{A}/\mathcal{B}) = I_\alpha(\mathcal{A})$.

Lemma 6.2: $\mathcal{A} \subset \mathcal{B} \Rightarrow \hat{I}_\alpha(\mathcal{A}/\mathcal{B}) = 0, \alpha > 0.$

Proof: Since $\chi(A) = 0$ or 1 at each point in Ω and $z_\alpha(0) = z_\alpha(1) = 0$, the conclusion follows from (6.3).

If $\alpha = 0$, then for all finite $\mathcal{A}$ and $\mathcal{B}$, it is obvious that $\hat{I}_0(\mathcal{A}/\mathcal{B}) = I_0(\mathcal{A})$ and hence Lemma 6.2 is not true for $\alpha = 0$ unless $\mathcal{A}$ is trivial.

Theorem 6.1. The conditional entropy $\hat{I}_\alpha(\mathcal{A}/\mathcal{B})$, $\alpha \geq 0$, is non-decreasing in its first argument and non-increasing in its second agrument.

Proof: $\mathcal{A}$ induces, on each $B \in \mathcal{B}$, a finite measurable partition $\mathcal{A}/B = \{B \cap A: A \in \mathcal{A}\}$. Defining $P(A/B) = \frac{\mu(A \cap B)}{\mu(B)}$, it is obvious that $\sum_{A \in \mathcal{A}} P(A/B) = 1$, so that each $B \in \mathcal{B}$ can be regarded as a space with measure unity, and the partition $\mathcal{A}/B$ has the entropy

$$(6.4) \qquad I_\alpha(\mathcal{A}/B) = \sum_{A \in \mathcal{A}} z_\alpha(P(A/B))$$

and from (6.3), $\hat{I}_\alpha(\mathcal{A}/\mathcal{B}) = \sum_{B \in \mathcal{B}} \mu(B) I_\alpha(\mathcal{A}/B)$.

Since $\mathcal{A} \subset \mathcal{B} \Rightarrow \mathcal{A}/C \subset \mathcal{B}/C \;\; \forall \; C \varepsilon \; \mathcal{C}$ where $\mathcal{C}$ is a finite partition, by theorem 4.3, $I_\alpha(\mathcal{A}/C) \leq I_\alpha(\mathcal{B}/C)$. Multiplying both sides by $\mu(C)$, and summing with respect to $C \; \varepsilon \; \mathcal{C}$, it follows that

(6.5) $$\mathcal{A} \subset \mathcal{B} \Rightarrow \hat{I}_\alpha(\mathcal{A}/\mathcal{C}) \leq \hat{I}_\alpha(\mathcal{B}/\mathcal{C}), \; \alpha \geq 0.$$

Thus $\hat{I}_\alpha(\mathcal{A}/\mathcal{B})$ is non-decreasing in its first argument. To prove second part of the theorem is equivalent to prove

(6.6) $$\mathcal{B} \subset \mathcal{C} \Rightarrow \hat{I}_\alpha(\mathcal{A}/\mathcal{C}) \leq \hat{I}_\alpha(\mathcal{B}/\mathcal{C}), \alpha > 0$$

Since $\mathcal{B} \subset \mathcal{C}$, therefore each $B \; \varepsilon \; \mathcal{B}$ is a disjoint union of elements of $\mathcal{C}$. Hence

$$\sum_{C \subset B} P(A/C)P(C/B) = \sum_{C \subset B} \frac{\mu(A \cap C)}{\mu(C)} \cdot \frac{\mu(B \cap C)}{\mu(B)} = P(A/B).$$

Since $z_\alpha(t)$ is a concave function of t, by Jenson's inequality [7],

(6.7) $$\sum_C P(C/B) z_\alpha(P(A/C) \leq z_\alpha(P(A/B))$$

$$\Rightarrow \sum_C \sum_B \sum_A \mu(B) P(C/B) \; z_\alpha \; (P(A/C) \leq \sum_A \; \sum_B \mu(B) z_\alpha(P(A/B))$$

$$\Rightarrow \sum_B \; \sum_C \; \mu(B \cap C) I_\alpha(\mathcal{A}/C) \leq \sum_B \mu(B) I_\alpha(\mathcal{A}/B)$$

$$\Rightarrow \hat{I}_\alpha(\mathcal{A}/\mathcal{C}) \leq \hat{I}_\alpha(\mathcal{A}/\mathcal{B}).$$

Remark 4: It is well-known [4] that $\hat{I}_1(\mathcal{A}/\mathcal{B})$ which is Shannon's conditional entropy, is non-decreasing in its first argument and non-increasing in its second argument. Theorem (6.1) shows that, in addition to $\hat{I}_1(\mathcal{A}/\mathcal{B})$, there are infinitely many conditional entropies which possess the same monotonic character with respect to its first and second arguments. If $\mathcal{B} = \{\Omega\}$, then (6.6) gives

(6.8) $$\hat{I}_\alpha(\mathcal{A}/\mathcal{C}) \leq \hat{I}_\alpha(\mathcal{A}), \; \alpha \geq 0,$$

equality in (6.8) being true when (i) both $\mathcal{A}$ and $\mathcal{C}$ are trivial or (ii) $\mathcal{A} \subset \mathcal{C}$ or (iii) $\alpha = 0$. Also, it is known that

$$(6.9) \qquad \hat{I}_1(\mathcal{A} \vee \mathcal{B}/\mathcal{C}) = \hat{I}_1(\mathcal{A}/\mathcal{C}) + \hat{I}_1(\mathcal{B}/\mathcal{A} \vee \mathcal{C}).$$

But

$$(6.10) \qquad \hat{I}_\alpha(\mathcal{A} \vee \mathcal{B}/\mathcal{C}) \neq \hat{I}_\alpha(\mathcal{A}/\mathcal{C}) + \hat{I}_\alpha(\mathcal{B}/\mathcal{A} \vee \mathcal{C}), \quad \alpha \geq 0,\ \alpha \neq 1, \quad \mathcal{A} \not\subset \mathcal{C},\ \mathcal{B} \not\subset \mathcal{C}.$$

Note that if (i) $\alpha = 0$, $\mathcal{A} = \mathcal{B} = \mathcal{C} = \{\Omega\}$ or (ii) $\mathcal{A} \subset \mathcal{C}$, $\mathcal{B} \subset \mathcal{C}$, we have

$$(6.11) \qquad \hat{I}_\alpha(\mathcal{A} \vee \mathcal{B}/\mathcal{C}) = \hat{I}_\alpha(\mathcal{A}/\mathcal{C}) + \hat{I}_\alpha(\mathcal{B}/\mathcal{A} \vee \mathcal{C}), \quad \alpha \geq 0,\ \alpha \neq 1$$

because in both cases, all the conditional entropies in (6.11) are zero. If $\mathcal{C}$ is trivial, then (6.9) reduces to

$$(6.12) \qquad I_1(\mathcal{A} \vee \mathcal{B}) = I_1(\mathcal{A}) + \hat{I}_1(\mathcal{B}/\mathcal{A}).$$

It may be noticed that sub-additivity of $I_1(\mathcal{A})$ is a consequence of (6.11) and the non-increasing nature of $\hat{I}_1(\mathcal{A}/\mathcal{B})$ w.r.t. its second argument, and since (6.11) is not true for all α, therefore $I_\alpha(\mathcal{A})$ need not be necessarily sub-additive for all α but there may exist values of α for which $I_\alpha(\mathcal{A})$ is sub-additive and this fact is obvious from (4.13).

From (4.1) it is clear that

$$(6.13) \qquad H_\alpha(\mathcal{A}) = (1-\alpha)^{-1} \log_2 \left[1 - (1 - 2^{1-\alpha}) I_\alpha(\mathcal{A})\right]$$

If we replace $I_\alpha(\mathcal{A})$ by $\hat{I}_\alpha(\mathcal{A}/\mathcal{B})$ on the R. H. S. in (6.13), we get one of Rényi's [10] conditional entropies, namely

$$\hat{H}_\alpha(\mathcal{A}/\mathcal{B}) = (1-\alpha)^{-1} \log_2 \left[1 - (1-2^{1-\alpha}) \hat{I}_\alpha(\mathcal{A}/\mathcal{B})\right], \alpha \geq 0, \alpha \neq 1$$

$$= (1-\alpha)^{-1} \log_2 \Big(\sum_{B \in \mathcal{B}} \mu(B) \sum_{A \in \mathcal{A}} p^\alpha(A/B) \Big)$$

$$= (1-\alpha)^{-1} \log_2 \Big(\sum_{A \in \mathcal{A}} \int_\Omega p\, (A/\tilde{\mathcal{B}})\, d\mu \Big)$$

We discuss the properties of $\hat{H}_\alpha(\mathcal{A}/\mathcal{B})$.

<u>Theorem 6.2</u>. If $\mathcal{A}$ and $\mathcal{B}$ are finite measureable partitions, then

(i) $\hat{H}_\alpha(\mathcal{A}/\mathcal{B}) = H_\alpha(\mathcal{A})$ if $\mathcal{A}$ and $\mathcal{B}$ are independent.

(ii) $\mathcal{A} \subset \mathcal{B} \Rightarrow \hat{H}_\alpha(\mathcal{A}/\mathcal{B}) = 0,\ \alpha > 0$

(iii) $\mathcal{A} \subset \mathcal{B} \Rightarrow \hat{H}_\alpha(\mathcal{A}/\mathcal{C}) \leq \hat{H}_\alpha(\mathcal{B}/\mathcal{C}), \alpha \geq 0$

(iv) $\mathcal{B} \subset \mathcal{C} \Rightarrow \hat{H}_\alpha(\mathcal{A}/\mathcal{C}) \leq \hat{H}_\alpha(\mathcal{A}/\mathcal{B}), \alpha \geq 0.$

Proof: (i) Since $\mathcal{A}$ and $\mathcal{B}$ are independent, therefore $\mu(A \cap B) = \mu(A)\mu(B) \quad \forall A \varepsilon \mathcal{A},\ B \varepsilon \mathcal{B}$. Hence $P(A/B) = \mu(A)$ and by (6.14) the result follows.

Note that $\mathcal{B} = \{\Omega\}$ is admissible.

(ii) The proof is similar to that of lemma 6.2

(iii) Since $\mathcal{A} \subset \mathcal{B}$, therefore each $A \varepsilon \mathcal{A}$ is a disjoint union of elements of $\mathcal{B}$. Let $A_i = \bigcup_j B_{i,j}$. Then for any $C \varepsilon \mathcal{C}$,

$$P(A_i/C) = P(\bigcup_j B_{i,j}/C) = \sum_j P(B_{i,j}/C).$$

Hence

$$(6.15) \qquad P^\alpha(A_i/C) = (\sum_j P(B_{i,j}/C))^\alpha \lesseqgtr \sum_j P^\alpha(B_{i,j}/C),\ \alpha \lessgtr 1,$$

so that

$$\sum_{C\varepsilon \mathcal{C}} \mu(C) \sum_{A\varepsilon \mathcal{A}} P^\alpha(A/C) \lesseqgtr \sum_{C\varepsilon \mathcal{C}} \mu(C) \sum_{B\varepsilon \mathcal{B}} P^\alpha(B/C),\ \alpha \lessgtr 1,$$

and from (6.14), after making some manipulations, the result follows.

(iv) It is obvious that $\mathcal{B} \subset \mathcal{C}$ implies

$$(6.16) \qquad \sum_C P(C/B)P^\alpha(A/C) \lesseqgtr (\sum_C P(C/B)P(A/C))^\alpha = P^\alpha(A/B),\ \alpha \lessgtr 1.$$

Multiplying both sides by $\mu(B)$ and summing w.r.t. A and B, we have

$$\sum_A \sum_C \mu(C)P^\alpha(A/C) \lesseqgtr \sum_A \sum_B \mu(B)P^\alpha(A/B), \alpha \lessgtr 1.$$

Taking logarithms and making necessary manipulations, the result follows immediately. For $\alpha = 1$, $\hat{H}_\alpha(\mathcal{A}/\mathcal{B}) = \hat{I}_1(\mathcal{A}/\mathcal{B})$ and (iii) and (iv) are well-known.

Corollary 1. If $\mathcal{B}$ is trivial, (iv) gives

(6.17) $\hat{H}_\alpha(\mathcal{A}/\mathcal{C}) \leq H_\alpha(\mathcal{A}), \alpha \geq 0$

Remark 5: It can be easily seen that

$$\hat{H}_\alpha(\mathcal{A} \vee \mathcal{B} / \mathcal{C}) = \hat{H}_\alpha(\mathcal{A}/\mathcal{C}) + \hat{H}_\alpha(\mathcal{B}/\mathcal{A} \vee \mathcal{C}),$$

only when (i) $\alpha = 0$, or (ii) $\mathcal{A} \subset \mathcal{C}$, $\mathcal{B} \subset \mathcal{C}$ (iii) $\alpha = 1$. Hence it can be easily concluded that $H_\alpha(\mathcal{A})$ is not sub-additive for $\alpha > 0, \alpha \neq 1$. When $\alpha = 0$ $H_0(\mathcal{A})$ is rather additive. When $\mathcal{A} \subset \mathcal{C}$, $\mathcal{B} \subset \mathcal{C}$, each of the conditional entropy under consideration reduces to zero. Also, for non-trivial partitions

(6.18) $H_\alpha(\mathcal{A} \vee \mathcal{B}) \neq H_\alpha(\mathcal{A}) + \hat{H}_\alpha(\mathcal{B}/\mathcal{A}), \ \alpha > 0, \ \alpha \neq 1.$

However, it is only for $\alpha = 0$ and 1, that $H_\alpha(\mathcal{A} \vee \mathcal{B}) = H_\alpha(\mathcal{A}) + \hat{H}_\alpha(\mathcal{B}/\mathcal{A})$ is true.

Let us define another conditional entropy

(6.19) $I_\alpha(\mathcal{A}/\mathcal{B}) = \sum_{A \varepsilon \mathcal{A}} \int_\Omega \mu^{\alpha-1}(\tilde{\mathcal{B}}) z_\alpha(P(A/\tilde{\mathcal{B}}))\, d\mu.$

Clearly

(6.20) $I_\alpha(\mathcal{A}/\mathcal{B}) = \sum_{B \varepsilon \mathcal{B}} \mu^\alpha(B) I_\alpha(\mathcal{A}/B).$

Obviously, if $\mathcal{B}$ is trivial, $I_\alpha(\mathcal{A}/\mathcal{B}) = I_\alpha(\mathcal{A})$. However, if $\mathcal{B}$ is not trivial, then even though $\mathcal{A}$ and $\mathcal{B}$ are independent, still $I_\alpha(\mathcal{A}/\mathcal{B}) \neq I_\alpha(\mathcal{A})$ except when $\alpha = 1$. Note that even for $\alpha = 0$, $I_\alpha(\mathcal{A}/\mathcal{B}) \neq I_\alpha(\mathcal{A})$. In this sense, $I_\alpha(\mathcal{A}/\mathcal{B})$ differs considerably from $\hat{I}_\alpha(\mathcal{A}/\mathcal{B})$. Also one can easily observe that, when $\mathcal{A}$ and $\mathcal{B}$ are independent,

(6.21) $I_\alpha(\mathcal{A}/\mathcal{B}) \leq I_\alpha(\mathcal{A}),\ \alpha \geq 1$

(6.22) $I_\alpha(\mathcal{A}/\mathcal{B}) \geq I_\alpha(\mathcal{A}),\ 0 \leq \alpha \leq 1.$

More precisely, if $\mathcal{A}$ and $\mathcal{B}$ are independent, then

(6.23) $I_\alpha(\mathcal{A}/\mathcal{B}) = M_\alpha(\mathcal{B})I_\alpha(\mathcal{A})$

and (6.21) and (6.22) are obvious conclusions of (6.23). It may be noticed that equality in (6.21) and (6.22) occurs when either (i) $\mathcal{B}$ is trivial or (ii) $\mathcal{A}$ is trivial or (iii) $\mathcal{A}$ and $\mathcal{B}$ are independent and $\alpha = 1$. In fact, for $\alpha \neq 1$,

$$I_\alpha(\mathcal{A}/\mathcal{B}) = I_\alpha(\mathcal{A}) \iff \mathcal{B} \text{ trivial},$$

From the physical point of view, (6.22) does not seem to be appealing because any prior knowledge must reduce our uncertainty. Hence, it is natural that $I_\alpha(\mathcal{A}/\mathcal{B})$ can serve as a useful definition of conditional entropy only for $\alpha \geq 1$. Moreover, from (6.23), it is now obvious that there do exist conditional entropies which do not necessarily reduce to marginal entropies when the partitions under consideration are independent.

<u>Theorem 6.3</u>: Let $\mathcal{A}$, $\mathcal{B}$, $\mathcal{C}$ be finite partitions. Then

(i) $\mathcal{A} \subset \mathcal{B} \Rightarrow I_\alpha(\mathcal{A}/\mathcal{B}) = 0,\ \alpha > 0.$

(ii) $\mathcal{B} \subset \mathcal{C} \Rightarrow I_\alpha(\mathcal{A}/\mathcal{C}) \leq I_\alpha(\mathcal{A}/\mathcal{B}),\ \alpha \geq 1,$

(iii) $I_\alpha(\mathcal{A} \vee \mathcal{B}/\mathcal{C}) = I_\alpha(\mathcal{A}/\mathcal{C}) + I_\alpha(\mathcal{B}/\mathcal{A} \vee \mathcal{C}),\ \alpha \geq 0.$

Proof: (i) The proof is similar to that of Lemma 6.2.

For $\alpha = 0$, (i) is true only when $\mathcal{A}$ is trivial.

(ii) It is obvious that, when $\mathcal{B} \subset \mathcal{C}$,

$$(6.7) \Rightarrow \sum_A \sum_B \sum_C \mu^\alpha(B)P(C/B)z_\alpha(P(A/C)) \leq \sum_A \sum_B \mu^\alpha(B)z_\alpha(P(A/B))$$

$$\Rightarrow \sum_A \sum_B \sum_C \mu^{\alpha}(B)P^{\alpha}(C/B)z_{\alpha}(P(A/C)) \leq I_{\alpha}(\mathcal{A}/\mathcal{B}),\ \alpha \geq 1$$

$$\Rightarrow \sum_A \sum_B \sum_C \mu^{\alpha}(B\ \ C)z_{\alpha}(P(A/C)) \leq I_{\alpha}(\mathcal{A}/\mathcal{B}), \alpha \geq 1$$

$$\Rightarrow I_{\alpha}(\mathcal{A}/\mathcal{C}) \leq I_{\alpha}(\mathcal{A}/\mathcal{B}), \text{ because } \sum_B \sum_C \mu^{\alpha}(B\cap C) = \sum_C \mu^{\alpha}(C),$$

Note that for $0 < \alpha < 1$, $I_{\alpha}(\mathcal{A}/\mathcal{B})$ is not non-increasing with respect to its second argument and hence, in this respect, it differs from $\hat{I}_{\alpha}(\mathcal{A}/\mathcal{B})$. Also, $\mathcal{B}\subset\mathcal{C} \Rightarrow I_{\alpha}(\mathcal{A}/\mathcal{C}) \geq I_{\alpha}(\mathcal{A}/\mathcal{B})$ when $\alpha = 0$.

(iii) By definition,

$$I_{\alpha}(\mathcal{A}\vee\mathcal{B}/\mathcal{C}) = \sum_{A\cap B\in\mathcal{A}\vee\mathcal{B}} \int_{\Omega} \mu^{\alpha-1}(\tilde{\mathcal{C}})z_{\alpha}(P(A\cap B/\tilde{\mathcal{C}}))d\mu$$

$$= \sum_{A\cap B} \sum_C \mu^{\alpha}(C)z_{\alpha}(P(A\cap B/C))$$

$$= \sum_A \sum_B \sum_C \mu^{\alpha}(C)z_{\alpha}\left(\frac{\mu(A\cap B\cap C)}{\mu(C)}\right)$$

$$= \left[\sum_A \sum_C \mu(A\cap C)\mu^{\alpha-1}(C) - \sum_A \sum_C \mu^{\alpha}(A\cap C)\right] c(\alpha)$$

$$+ \left[\sum_A \sum_C \mu^{\alpha}(A\cap C) - \sum_A \sum_B \sum_C \mu^{\alpha}(A\cap B\cap C)\right] c(\alpha)$$

$$= I_{\alpha}(\mathcal{A}/\mathcal{C}) + \sum_{B\in\mathcal{B}} \int_{\Omega} \mu^{\alpha-1}(A\cap C)z_{\alpha}\left(\frac{P(B/A\cap C)}{\mu(A\cap C)}\right) d\mu$$

$$= I_{\alpha}(\mathcal{A}/\mathcal{C}) + I_{\alpha}(\mathcal{B}/\mathcal{A}\vee\mathcal{C}).$$

<u>Corollary 1</u>: If $\mathcal{C}$ is trivial, then (iii) gives

(6.24) $\quad I_{\alpha}(\mathcal{A}\vee\mathcal{B}) = I_{\alpha}(\mathcal{A}) + I_{\alpha}(\mathcal{B}/\mathcal{A}),\ \alpha \geq 0.$

<u>Corollary 2</u>: If $\mathcal{A}\subset\mathcal{B}$, then $\mathcal{A}\vee\mathcal{B} = \mathcal{B}$. Hence (6.24) gives

(6.25) $\quad I_{\alpha}(\mathcal{B}/\mathcal{A}) = I_{\alpha}(\mathcal{B}) - I_{\alpha}(\mathcal{A}),\ \alpha \geq 0.$

In other words, if $\mathcal{B}$ is a refinement of $\mathcal{A}$, then the difference $I_\alpha(\mathcal{B}) - I_\alpha(\mathcal{A})$, which is always non-negative, is precisely the conditional entropy $I_\alpha(\mathcal{B}/\mathcal{A})$. But $I_\alpha(\mathcal{B}/\mathcal{A}) \geq 0 \not\Rightarrow \mathcal{A} \subset \mathcal{B}$.

<u>Corollary 3</u>: If $\mathcal{A}$ and $\mathcal{B}$ are independent, then from (6.24) and (6.20) we get

$$(6.26) \qquad I_\alpha(\mathcal{A} \vee \mathcal{B}) = I_\alpha(\mathcal{A}) + M_\alpha(\mathcal{A}) I_\alpha(\mathcal{B}), \quad \alpha \geq 0$$

which is equivalent to (4.10) and (4.11), for $\alpha > 0$.

<u>Corollary 4</u>: If (i) $\mathcal{A}$ and $\mathcal{C}$ are independent, (ii) $\mathcal{B}$ and $\mathcal{A} \vee \mathcal{C}$ are independent (iii) $\mathcal{A} \vee \mathcal{B}$ and $\mathcal{C}$ are independent, then

$$I_\alpha(\mathcal{A} \vee \mathcal{B} / \mathcal{C}) = M_\alpha(\mathcal{C}) I_\alpha(\mathcal{A} \vee \mathcal{B}), \quad I_\alpha(\mathcal{A}/\mathcal{C}) = M_\alpha(\mathcal{C}) I_\alpha(\mathcal{A})$$

$$I_\alpha(\mathcal{B} / \mathcal{A} \vee \mathcal{C}) = M_\alpha(\mathcal{A} \vee \mathcal{C}) I_\alpha(\mathcal{B}).$$

Hence, from theorem (6.3) (iii), (6.26) again follows. But it should be noticed that we have neither assumed $\mathcal{C}$ to be trivial nor $\mathcal{A}$ and $\mathcal{B}$ to be independent. Rather we can conclude that $\mathcal{A}$ and $\mathcal{B}$ independent. From our assumptions, (i) and (ii) it is obvious that

$$\mu(A \cap B \cap C) = \mu(B)\mu(A \cap C) = \mu(A)\mu(B)\mu(C).$$

Also, from (iii), $\mu(A \cap B \cap C) = \mu(A \cap B).\mu(C)$. Thus $\mu(A \cap B) = \mu(A).\mu(B)$ $\forall A \in \mathcal{A}$, $B \in \mathcal{B}$. Hence $\mathcal{A}$ and $\mathcal{B}$ are independent and that is why we have got (6.26) again. Note that $\mathcal{B}$ and $\mathcal{C}$ may no longer be independent.

<u>Corollary 5</u>: If $\mathcal{A} \subset \mathcal{B}$, then $\mathcal{A} \vee \mathcal{B} = \mathcal{B}$ and using non-negativity of $I_\alpha(\mathcal{B}/\mathcal{A} \vee \mathcal{C})$, it follows from theorem (6.3) (iii), that

$$(6.27) \qquad \mathcal{A} \subset \mathcal{B} \Rightarrow I_\alpha(\mathcal{A}/\mathcal{C}) \leq I_\alpha(\mathcal{B}/\mathcal{C}) \quad \alpha \geq 0,$$

that is, $I_\alpha(\mathcal{A}/\mathcal{B})$ is non-decreasing function of its first argument for all $\alpha \geqslant 0$.

Remark 6: It is clear that

$$(6.28) \quad \hat{I}_\alpha(\mathcal{A}/\mathcal{B}) \leq I_\alpha(\mathcal{A}/\mathcal{B}),\ 0 \leq \alpha \leq 1,$$

$$\hat{I}_\alpha(\mathcal{A}/\mathcal{B}) \geqslant I_\alpha(\mathcal{A}/\mathcal{B}),\ 1 \leq \alpha < \infty,$$

equality in (6.28) and (6.29) occurs when (i) $\mathcal{A} \subset \mathcal{B}$ or (ii) $\alpha = 1$, $\mathcal{A}$ and $\mathcal{B}$ being any finite partitions or (iii) $\mathcal{B}$ is trivial or (iv) $\alpha = 0$, at least one of $\mathcal{A}$ and $\mathcal{B}$ is trivial. Hence it is obvious that

$$(6.30) \quad I_\alpha(\mathcal{A} \vee \mathcal{B}) \geqslant I_\alpha(\mathcal{B}) + \hat{I}_\alpha(\mathcal{A}/\mathcal{B}),\ 0 < \alpha < 1$$

$$I_\alpha(\mathcal{A} \vee \mathcal{B}) \leq I_\alpha(\mathcal{B}) + \hat{I}_\alpha(\mathcal{A}/\mathcal{B}),\ 1 \leq \alpha < \infty.$$

Finally,

$$I_\alpha(\mathcal{A} \vee \mathcal{B}/\mathcal{C}) > \max\left\{ I_\alpha(\mathcal{A}/\mathcal{C}),\ I_\alpha(\mathcal{B}/\mathcal{C})\right\} \quad ,\alpha \geqslant 0.$$

From (6.13), the corresponding conditional entropy is

$$(6.32) \quad H_\alpha(\mathcal{A}/\mathcal{B}) = 1-\alpha)^{-1} \log_2 \left[1 - M_\alpha(\mathcal{B}) + M_\alpha(\mathcal{A} \vee \mathcal{B})\right] \quad ,\alpha \geqslant 0, \alpha \neq 1.$$

which has, of course, not been proposed by Rényi [10]. If $\mathcal{B}$ is trivial, $H_\alpha(\mathcal{A}/\mathcal{B}) = H_\alpha(\mathcal{A})$, but if $\mathcal{A}$ and $\mathcal{B}$ are independent, $H_\alpha(\mathcal{A}/\mathcal{B}) \neq H_\alpha(\mathcal{A})$. In other words there are conditional entropies of which Rényi's entropy is a particular case but which do not reduce to Rényi's entropy when the partitions under consideration are independent. A detailed discussion of $H_\alpha(\mathcal{A}/\mathcal{B})$ shall be presented elsewhere. It is appropriate to remark that

$$(6.33) \quad H_\alpha(\mathcal{A} \vee \mathcal{B}) \neq H_\alpha(\mathcal{A}) + H_\alpha(\mathcal{B}/\mathcal{A}),\ \alpha > 0,\ \alpha \neq 1.$$

However $H_\alpha(\mathcal{A} \vee \mathcal{B}) = H_\alpha(\mathcal{A}) + H_\alpha(\mathcal{B}/\mathcal{A})$ holds only when (i) $\alpha = 0$, at least one of $\mathcal{A}$ and $\mathcal{B}$ is trivial or (ii) $\alpha = 1$.

Lastly, we define another conditional entropy

$$(6.34)\qquad \bar{I}_\alpha(\mathcal{A}/\mathcal{B}) = \sum_{A\in\mathcal{A}} \int_\Omega \frac{\mu^{\alpha-1}(\tilde{\mathcal{B}})}{M_\alpha(\mathcal{B})}\, z_\alpha(PA/\tilde{\mathcal{B}})\,d\mu .$$

Obviously

$$\bar{I}_\alpha(\mathcal{A}/\mathcal{B}) = c(\alpha)\left[1 - \frac{M_\alpha(\mathcal{A}\vee\mathcal{B})}{M_\alpha(\mathcal{B})}\right] = \left[M_\alpha(\mathcal{B})\right]^{-1} I_\alpha(\mathcal{A}/\mathcal{B})$$

so that

$$(6.35)\qquad I_\alpha(\mathcal{A}/\mathcal{B}) = M_\alpha(\mathcal{B})\bar{I}_\alpha(\mathcal{A}/\mathcal{B}),\ \alpha > 0$$

Hence the properties of $\bar{I}_\alpha(\mathcal{A}/\mathcal{B})$ can be easily derived from those of $I_\alpha(\mathcal{A}/\mathcal{B})$. We enumerate some of its properties

(A_1) $\mathcal{A}\subset\mathcal{B} \Rightarrow \bar{I}_\alpha(\mathcal{A}/\mathcal{B}) = 0$

(A_2) $\mathcal{B}\subset\mathcal{C} \Rightarrow \bar{I}_\alpha(\mathcal{A}/\mathcal{C}) \leq M_\alpha(\mathcal{B})\left[M_\alpha(\mathcal{C})\right]^{-1}\bar{I}_\alpha(\mathcal{A}/\mathcal{B}), \alpha \geq 1$

and if $\mathcal{B}$ is trivial, then

$$\bar{I}_\alpha(\mathcal{A}/\mathcal{C}) \leq \left[M_\alpha(\mathcal{C})\right]^{-1} I_\alpha(\mathcal{A}),\quad \alpha \geq 1.$$

On the other hand if $\mathcal{A}$ and $\mathcal{B}$ are independent, then from (6.35),

$$\bar{I}_\alpha(\mathcal{A}/\mathcal{B}) = I_\alpha(\mathcal{A}),\quad \alpha \geq 0.$$

But $\bar{I}_\alpha(\mathcal{A}/\mathcal{B}) \leq I_\alpha(\mathcal{A})$ is not, in general, true.

(A_3) $\bar{I}_\alpha(\mathcal{A}/\mathcal{B})$ is non-decreasing in its first argument.

(A_4) $\bar{I}_\alpha(\mathcal{A}/\mathcal{B})$ is not non-increasing in its second argument.

(A_5) $\bar{I}_\alpha(\mathcal{A}\vee\mathcal{B}/\mathcal{C}) = \bar{I}_\alpha(\mathcal{A}/\mathcal{C}) + \dfrac{M_\alpha(\mathcal{A}\vee\mathcal{C})}{M_\alpha(\mathcal{C})}\,\bar{I}_\alpha(\mathcal{B}/\mathcal{A}\vee\mathcal{C})$,

and if $\mathcal{C}$ is trivial, then

$$(6.36)\qquad I_\alpha(\mathcal{A}\vee\mathcal{B}) = I_\alpha(\mathcal{A}) + M_\alpha(\mathcal{A})\,\bar{I}_\alpha(\mathcal{B}/\mathcal{A}).$$

Rényi's conditional entropy corresponding to $\bar{I}_\alpha(\mathcal{A}/\mathcal{B})$ is

$$(6.37)\qquad \bar{H}_\alpha(\mathcal{A}/\mathcal{B}) = (1-\alpha)^{-1}\log_2\Big(\sum_B \Big(\frac{\mu^\alpha(B)}{\sum_B \mu^\alpha(B)}\Big)\sum_A p^\alpha(A/B)\Big),\ \alpha\geq 0, \alpha\neq 1$$

$$= (1-\alpha)^{-1} \log_2 \sum_{A \in \mathcal{A}} \int_\Omega \frac{\mu^{\alpha-1}(\tilde{\mathcal{B}})}{\int_\Omega \mu^\alpha(\tilde{\mathcal{B}})d\mu} p^\alpha(A/\tilde{\mathcal{B}})d\mu, \alpha \geq 0, \alpha \neq 1.$$

We state the following theorem without proof.

Theorem 6.4: Let $\mathcal{A}$, $\mathcal{B}$, $\mathcal{C}$ be finite partitions. Then, for $\alpha > 0$,

(B_1) $\bar{H}_\alpha(\mathcal{A}/\mathcal{B}) = H_\alpha(\mathcal{A}) \Leftrightarrow$ $\mathcal{A}$ and $\mathcal{B}$ are independent.

(B_2) $\mathcal{A} \subset \mathcal{B} \Rightarrow \bar{H}_\alpha(\mathcal{A}/\mathcal{B}) = 0$.

(B_3) $\bar{H}_\alpha(\mathcal{A} \vee \mathcal{B}/\mathcal{C}) = \bar{H}_\alpha(\mathcal{A}/\mathcal{C}) + \bar{H}_\alpha(\mathcal{B}/\mathcal{A} \vee \mathcal{C})$.

(B_4) $\bar{H}_\alpha(\mathcal{A}/\mathcal{B})$ is non-decreasing in its first argument.

Remark 7: If $\alpha = 0$, $\bar{H}_\alpha(\mathcal{A}/\mathcal{B}) = H_\alpha(\mathcal{A})$ for all $\mathcal{A}$ and $\mathcal{B}$. Hence $\bar{H}_0(\mathcal{A}/\mathcal{B}) = H_0(\mathcal{A})$ does not necessarily imply that $\mathcal{A}$ and $\mathcal{B}$ are independent. Since $H_0(\mathcal{A}) = 0 \Leftrightarrow \mathcal{A}$ must be trivial, it follows that (B_2) is true for $\alpha = 0$ only when $\mathcal{A} = \{\Omega\}$. (B_4) remains true even for $\alpha = 0$. (B_3) also remains valid for $\alpha = 0$. An obvious deduction from (B_4) is

(6.38) $H_\alpha(\mathcal{A} \vee \mathcal{B}) = H_\alpha(\mathcal{A}) + \bar{H}_\alpha(\mathcal{B}/\mathcal{A})$, $\alpha > 0$,

but $\bar{H}_\alpha(\mathcal{A}/\mathcal{B}) \leq H_\alpha(\mathcal{A})$ is not necessarily true except for $\alpha = 1$ and 0, equality holding in the latter case. If $\mathcal{A} \subset \mathcal{B}$, then $\bar{H}_\alpha(\mathcal{B}/\mathcal{A})$ is precisely the difference $H_\alpha(\mathcal{B}) - H_\alpha(\mathcal{A})$ which is non-negative.

Since $\bar{H}_\alpha(\mathcal{A}/\mathcal{B}) \leq H_\alpha(\mathcal{A})$ is true only for $\alpha = 0$ and 1, it is clear that $H_\alpha(\mathcal{A})$ is sub-additive only for $\alpha = 1$ and additive for $\alpha = 0$. In this sense, $I_\alpha(\mathcal{A})$ seems to be a more useful generalization of $I_1(\mathcal{A})$ i.e. Shannon's entropy.

The above proposed definitions of conditional entropies can be used to study the entropy of a finite partition with respect to a measure-preserving transformation. We hope to study such problems in our subsequent work and discuss the roles of these entropies in ergodic theory. The entropies defined and studied here will also serve as a basis in proving Shannon-Wolfowitz coding theorem for various channels.

The authors are thankful to Professor I. Z. Chorneyko for fruitful discussions. The authors are indebted to Professor J. Aczél for rereading the manuscript and offering valuable comments and suggestions.

References

[1] Aczél, J. (1966), Lectures on Functional equations and their applications, Academic Press, New York.

[2] Arnold, V. I. and Avez, A., (1968), Ergodic Problems of Classical Mechanics, W. A. Benjamin, Inc., New York.

[3] Behara, M. (1968), "Entropy as a Measure of Utility in Decision Theory. Invited paper presented at Mathematisches Institut, Oberwolfach, W. Germany.

[4] Billingsley, P. (1965), Ergodic Theory and Information, John Wiley and Sons, Inc., New York.

[5] Brown, T. A. (1963), Entropy and Conjugacy, Ann. Math. Stat., 34, 226-232.

[6] Chaundy, T. W.; Mcleod, J. B. (1960), "On a Functional equation", Proc. Edin. Math. Soc. Notes, 43 7-8.

[7] Hardy, G. H.; Littlewood, J. E.; Polya, G. (1952), Inequalities, London, Cambridge Univ. Press.

[8] Meshalkin, L. D. (1959), "A case of isomorphism of Bernoulli schemes", Dokl. Akad. Nauk. SSSR, 128, 41-44

[9] Renyi, A. (1961) "On Measures of Entropy and Information", Proc. Fourth Berk. Symp. on Math. Stat. and Prob. 1, 547-561.

[10] Renyi, A. (1960), "Az informacióelmélét Nehány alepvető kerdése", A Mag. Tud. Akad. III (Mat. es Fiz), Ostal. kozl. 257-282.

[11] Shannon, C. E. (1948), "A Mathematical theory of communication", Bell Sys. Tech. Jour., 27; 379-423, 623-656.

ON DISCRIMINANT DECISION FUNCTION IN COMPLEX GAUSSIAN DISTRIBUTIONS

N. Giri
University of Montreal

0. Introduction and summary: Let $\xi = (\xi_1,\ldots,\xi_p)'$, $\eta = (\eta_1,\ldots,\eta_p)'$ be two p-dimensional independent complex Gaussian random variables with complex means $E(\xi) = \alpha = (\alpha_1,\ldots,\alpha_p)'$, $E(\eta) = \beta = (\beta_1,\ldots \beta_p)$ and with the same Hermitian positive definite complex covariance matrix $\Sigma = E(\xi-\alpha)(\xi-\alpha)^* = E(\eta-\beta)(\eta-\beta)^*$ where $(\xi-\alpha)^*$, $(\eta-\beta)^*$ are the adjoints of $(\xi-\alpha)$, $(\eta-\beta)$ respectively. The probability density functions of ξ and η (with respect to the Lebesgue measure in the 2p dimensional Euclidean spaces of real and imaginary parts of ξ, η respectively) are given by

$$(0.1)\qquad \Pi_1 = p(\xi|\alpha,\Sigma) = \pi^{-p}(\det \Sigma)^{-1} \exp\{-(\xi-\alpha)^* \ \Sigma^{-1}(\xi-\alpha)\},$$

$$\Pi_2 = p(\eta|\alpha,\Sigma) = \pi^{-p}(\det \Sigma)^{-1} \exp\{-(\eta-\beta)^* \ \Sigma^{-1}(\eta-\beta)\},$$

with $E(\xi-\alpha)(\xi-\alpha)' = E(\eta-\beta)(\eta-\beta)' = 0$. In classifying an observation $Z = (Z_1,\ldots,Z_p)'$ either to Π_1 or to Π_2 one considers, as in the real case, the ratio $\Pi_2|\Pi_1$ and classify Z to

$$(0.2)\qquad \begin{array}{ll} \Pi_2; & \text{if } \ p(Z|\beta,\Sigma) \,/\, p(Z|\alpha,\Sigma) \geq K \\ \Pi_1; & \text{if otherwise.} \end{array}$$

The constant K depends on the size of the errors of misclassification.

Now,

$$(0.3) \quad p(Z|\beta,\Sigma)/p(Z|\alpha,\Sigma) = \exp\{U\}$$

where $U = Z^* \Sigma^{-1}(\alpha-\beta) + (\alpha-\beta)^* \Sigma^{-1}Z - \alpha^*\Sigma^{-1}\alpha + \beta^*\Sigma^{-1}\beta$. It may be verified that

$$(0.4) \quad E_{\Pi_1}(U) = (\alpha-\beta)^* \Sigma^{-1}(\alpha-\beta), \; E_{\Pi_2}(U) = -(\alpha-\beta)^* \Sigma^{-1}(\alpha-\beta).$$

The complex covariance matrix of U is $2(\alpha-\beta)^*\Sigma^{-1}(\alpha-\beta)$. Furthermore, as in the real case, $Z^*\Sigma^{-1}(\alpha-\beta)$ $\left((\alpha-\beta)^* \Sigma^{-1}Z\right)$ is the linear function which maximized

$$(0.5) \quad \left|E_{\Pi_1}(Z^*d) - E_{\Pi_2}(Z^*d)\right|^2/E_{\Pi_1}(Z^*d - \alpha^*d)^*(Z^*d - \alpha^*d)$$

$$= (X'(\alpha-\beta)(\alpha-\beta)^*X + Y'(\alpha-\beta)(\alpha-\beta)^*Y)/X'\Sigma X + Y/\Sigma Y).$$

for all choices of X, Y where $d = X + iY$ and subject to the condition that $X'\Sigma X + Y'\Sigma Y$ is constant. Using Lagrange multiplier λ and differenciating with respect to X and Y we get

$$(0.6) \quad \begin{aligned} (\alpha-\beta)(\alpha-\beta)^*X &= \lambda\Sigma X \\ (\alpha-\beta)(\alpha-\beta)^*Y &= \lambda\Sigma Y \end{aligned}$$

or equivalently d is proportional to $\Sigma^{-1}(\alpha-\beta)$.

We will consider here tests of hypothesis concerning $\Sigma^{-1}(\alpha-\beta)$. Since $\xi-\eta$ is also p-variate complex Gaussian, for simplicity, we will consider testing problems concerning $\Sigma^{-1}\alpha$ where ξ is p-variate complex Gaussian with complex mean α and with Hermitian positive definite covariance matrix Σ. Let $\Sigma^{-1}\alpha = \Gamma = (\Gamma_1,\ldots,\Gamma_p)'$. We will consider here the following problems.

A. To test the null hypothesis $H_{10}: \Gamma_{p_1+1} = \ldots = \Gamma_p = 0$ against the alternatives $H_{11}: \Gamma \neq 0$ when both α,Σ are unknown and $p_1 < p$;

B. To test the null hypothesis $H_{20}: \Gamma_{p_1+1} = \ldots = \Gamma_p = 0$ against the alternatives $H_{21}: \Gamma_{p_1+p_2+1} = \ldots = \Gamma_p = 0$ when both α, Σ are unknown and $p_1 + p_2 < p$;

C. To test the null hypothesis $H_{30}: \Gamma = 0$ against the alternatives $H_{31}: \Gamma_{p_1+1} = \ldots = \Gamma_p = 0$ when α, Σ are both unknown and $p_1 < p$;

D. To test the null hypothesis $H_{40}: \Gamma = 0$ against the alternatives $H_{41}: \Gamma \neq 0$ when α, Σ are both unknown.

The problem D is equivalent to the problem of testing $\alpha = 0$ against the alternatives $\alpha \neq 0$. the likelihood ratio test of this problem has been obtained in Giri (1965). This test is the complex analogue of Hotelling's T^2-test in the real case and it possesses all of the optimum properties possessed by the T^2-test in the real case (see Giri (1965), Salaevskii (1968)).

We will find the likelihood ratio test for each of these problems and study some of their optimum properties. As in the real case, the importance of this problem is self evident.

Let $\xi_1, \ldots, \xi_N$ be N independent observation on ξ. The likelihood of $\xi_1, \ldots, \xi_N$ is given by

$$(0.7) \qquad L(\alpha, \Sigma) = \Pi^{-Np} (\det \Sigma)^{-N} \exp \{-\sum_1^N (\xi_i - \alpha)^* \Sigma^{-1} (\xi_i - \alpha)\}.$$

Write $\bar{\xi} = \sum_1^N \xi_i / N$, $S = \sum_1^N (\xi_i - \bar{\xi})(\xi_i - \bar{\xi})^*$. We will assume throughout that $N > p$, so that S is positive definite Hermitian with probability one. From Giri (1965) it follows that $\bar{\xi}$ is independent of S in distribution and S is complex Wishart with parameters $(N-1, p, \Sigma)$, i.e. its probability density function with respect to the Lebesque measure $[\prod_{i>j} dS_{ijR}\, dS_{ijI} \prod_i dS_{ii}]$ (S_{ij} being the elements of S and $S_{ij} = S_{ijR} + iS_{ijI}$) is given by

$$(0.8) \quad p(S) = C(\det \Sigma^{-1})^{N-1} (\det S)^{N-p-1} \exp(-\mathrm{tr}\, \Sigma^{-1} S),$$

where $C^{-1} = \pi^{\frac{p(p-1)}{2}} \Gamma(N) \ldots \Gamma(N-p+1)$; and $\sqrt{N}\,\overline{\xi}$ is complex Gaussian with mean $\sqrt{N}\,\alpha$ and Hermitian positive definite covariance matrix Σ. Furthermore $(\sqrt{N}\,\overline{\xi}, S)$ is jointly sufficient for (α, Σ).

1. An Important Distribution: For any complex p-vector $b = (b_1, \ldots, b_p)$, and for any complex $p \times p$ matrix A we write

$$b = (b_{(1)}, \ldots, b_{(k)}), \quad b_{[i]} = (b_{(1)}, \ldots, b_{(i)})$$

$$(1.1) \quad A = \begin{pmatrix} A_{(11)}, \ldots, A_{(1k)} \\ \cdots\cdots\cdots \\ A_{(k1)}, \ldots, A_{(kk)} \end{pmatrix}, \quad A_{[ii]} = \begin{pmatrix} A_{(11)}, \ldots, A_{(1i)} \\ \cdots\cdots\cdots \\ A_{(i1)}, \ldots, A_{(ii)} \end{pmatrix}$$

where $b_{(i)}$ is the $p_i \times 1$ subvector of b and $A_{(ii)}$ is the $p_i \times p_i$ submatrix of A and $\sum_1^k p_i = p$. Let us now define $R_1, \ldots, R_k$; $\delta_1, \ldots, \delta_k$ by

$$(1.2) \quad \sum_1^i R_j = N\,\overline{\xi}^*_{[i]} (S_{[ii]} + N\,\overline{\xi}_{[i]} \overline{\xi}^*_{[i]})^{-1} \overline{\xi}_{[i]}$$

$$= N\,\overline{\xi}^*_{[i]} S^{-1}_{[ii]} \overline{\xi}_{[i]} / (1 + N\,\overline{\xi}^*_{[i]} S^{-1}_{[ii]} \overline{\xi}_{[i]}),$$

$$\sum_1^i \delta_j = N\,\alpha^*_{[i]} \Sigma^{-1}_{[ii]} \alpha_{[i]}, \quad i = 1, \ldots, k.$$

From Giri (1970) the joint distribution of $R_1, \ldots, R_K$ is given by

$$(1.3) \quad p(R_1, \ldots, R_k) = \Gamma(N) \left[\Gamma(N - \sum_1^k p_i) \prod_1^k \Gamma(p_i) \right]^{-1}$$

$$(1 - \sum_1^k R_j)^{N - \sum_1^k p_j - 1} \prod_1^k R_i^{p_i - 1}$$

$$. \exp \{- \sum_1^k \delta_i + \sum_1^k R_j \sum_{i>j} \delta_i\} \prod_1^k \phi(N - \sigma_{i-1}, p_i; R_i \delta_i)$$

where $\sigma_i = \sum_1^i p_j$, $\sigma_o = 0$ and ϕ is the confluent hypergeometric function given by

(1.4) $\phi(a,b;x) = 1 + \frac{a}{b} x + \frac{a(a+1)}{b(b+1)} \frac{x^2}{2!} + \dots$.

Furthermore the marginal distribution of $R_1,\dots,R_j$ $(j < k)$ is obtained from (1.3) by replacing k by J. (In this section the notation Γ is used for Gamma function).

2. <u>The likelihood ratio tests:</u>

<u>A</u>. Here $k = 2$, $p_1 + p_2 = p$. In terms of $\Gamma(=\Sigma^{-1}\alpha)$ and Σ the likelihood of $\xi_1,\dots,\xi_N$ is given by

(2.1) $$L(\Gamma,\Sigma) = \pi^{-Np}(\det \Sigma)^{-N} \exp \{-tr(\Sigma^{-1} (S + N \bar{\xi} \bar{\xi}^*) - N \bar{\xi} \Gamma^* - N \Gamma \bar{\xi}^* + N\Sigma\Gamma\Gamma^*)\} .$$

Under H_{10}; $\Gamma_{p_1+1} = ,\dots, = \Gamma_p = 0$ and under H_{11}: $\Gamma \neq 0$. Now

(2.2) $$\max_{H_{10}} L(\Gamma,\Sigma)$$

$$= \max_{\Gamma_{(1)}, \Sigma} \pi^{-Np}(\det \Sigma)^{-N} \exp (- tr (\Sigma^{-1}(S + N \bar{\xi} \bar{\xi}^*) - N \Gamma_{(1)} \bar{\xi}^*_{(1)} - N \bar{\xi}_{(1)} \Gamma^*_{(1)} + N \Sigma_{(11)} \Gamma_{(1)} \Gamma^*_{(1)})),$$

$$= \max_{\Sigma} \; \Pi^{-Np}(\det(\Sigma))^{-N} \; \exp\left(-tr(\Sigma^{-1}(S + N\,\bar{\xi}\,\bar{\xi}^*)\right.$$

$$\left. - N\,\Sigma^{-1}_{(11)}\;\bar{\xi}_{(1)}\;\bar{\xi}^*_{(1)}\;)\right).$$

Since Σ and $S + N\,\bar{\xi}\,\bar{\xi}^*$ are Hermitian positive definite, there exists complex upper traingular matrices K and T such that

(2.3) $\quad S + N\,\bar{\xi}\,\bar{\xi}^* = TT^*, \qquad \Sigma = KK^*,$

Let,

(2.4) $$T = \begin{pmatrix} T_{(11)} & T_{(12} \\ 0 & T_{(22)} \end{pmatrix}, \quad K = \begin{pmatrix} K_{(11)} & K_{(12)} \\ 0 & K_{(22)} \end{pmatrix}$$

Then

(2.5) $$T^{-1} = \begin{pmatrix} T^{-1}_{(11)} & -(T^{-1}_{(11)}\;T_{(12)}\;T^{-1}_{(22)}) \\ 0 & T^{-1}_{(22)} \end{pmatrix}$$

$$K^{-1} = \begin{pmatrix} K^{-1}_{(11)} & -(K^{-1}_{(11)}\;K_{(12)}\;K^{-1}_{(22)}) \\ 0 & K^{-1}_{(22)} \end{pmatrix}$$

and $K_{(11)}\;K^*_{(11)} = \Sigma_{(11)}$, $T_{(11)}\;T^*_{(11)} = S_{(11)}$.

Let $L = T^{-1}K$ and $Z_{(1)} = T^{-1}_{(11)}\;\bar{\xi}_{(1)}$, we get from (2.2)

(2.6) $$\max_{H_{10}} L(\Gamma, \Sigma) = \max_{L} \; \Pi^{-Np}\,(\det\,(S + N\;\bar{\xi}\;\bar{\xi}^*))^{-N}(\det(LL^*))^{-N}$$

$$\exp \{- \operatorname{tr}(LL^*)^{-1} - N(L_{(11)} L^*_{(11)})^{-1} Z_{(1)} Z^*_{(1)}\}$$

$$= \max_{\Delta} \Pi^{-Np} (\det(S + N \bar{\xi} \bar{\xi}^*))^{-N}$$

$$(\det(\Delta_{(22)})(\Delta_{(11)} - \Delta_{(12)} \Delta^{-1}_{(22)} \Delta_{(21)}))^N$$

$$\exp (-\operatorname{tr}(\Delta_{(11)} + \Delta_{(22)} - (\Delta_{(11)} - \Delta_{(12)} \Delta^{-1}_{(22)} \Delta_{(21)})NZ_{(1)} Z^*_{(1)}))$$

$$= (N\Gamma)^{-Np} (\det(S+N \bar{\xi} \bar{\xi}^*))^{-N} (\det(I-NZ_{(1)} Z^*_{(1)}))^{-N}$$

$$. \exp (-Np)$$

$$= (N\Pi)^{-Np} (\det(S + N \bar{\xi} \bar{\xi}^*))^{-N} (1-N \bar{\xi}^*_{(1)} (S_{(11)} + N \bar{\xi}_{(1)} \bar{\xi}^*_{(1)})^{-1} \bar{\xi}_{(1)})^{-N} . \exp (-Np).$$

Hence,

$$(2.7) \qquad \lambda = \max_{H_{10}} L(\Gamma,\Sigma) / \max_{H_{11}} L(\Gamma, \Sigma)$$

$$= ((1 - R_1 - R_2)(1 - R_1)^{-1})^N .$$

Thus the likelihood ratio test rejects H_{10} for small values of $Z = (1- R_1 - R_2)(1 - R_1)^{-1}$. From (1.3) the joint distribution of Z and R_1 under H_{10} is given by

$$\exp(-\delta_1) \; Z^{N-p_1-p_2-1} (1-Z)^{p_2-1} \; [B(N-p_1-p_2, p_2)]^{-1}$$

$$\sum_{j=0}^{\infty} \frac{(\delta_1 R_1)^j R_1^{p_1-1} (1-R_1)^{N-p_1-1}}{j!\, B(N-p_1,\, p_1+j)} .$$

Thus under H_{10}, Z is independent of R_1 and Z is distributed as beta with parameters $(N - p_1 - p_2, p_2)$.

B. Here $k = 3$, $p_1 + p_2 + p_3 = p$. Proceeding exactly in the same way as in A, it can be shown that the likelihood ratio test of H_{20} against H_{21} rejects H_{20} for small values of Z and under H_{20}, Z is distributed as central beta with parameter $(N - p_1 - p_2, p_2)$.

C. Here $k = 2$, $p_1 + p_2 = p$. Here also, it is evident from the computations of case A that the likelihood ratio test rejects H_{30} for small values of $1 - R_1$ and under H_{30} (from (1.3)) $1 - R_1$ is distributed as central beta with parameters $(N - p_1, p_1)$.

3. Optimum Properties: Let G be a group of nonsingular complex $p \times p$ matrices of the following form, $g \in G$

$$g = \begin{pmatrix} g_{(11)} & 0 & 0 & 0 \\ g_{(21)} & g_{(22)} & 0 & 0 \\ 0 & 0 & 0 & 0 \\ g_{(k1)} & 0 & 0 & g_{(kk)} \end{pmatrix} \qquad (3.1)$$

where, as before, $g_{(ii)}$ is the $p_i \times p_i$ submatrix of g and $\sum_i^k p_i = p$.

Let us now consider the group of transformations G with transforms. $(\bar{\xi}, S; \alpha, \Sigma) \longrightarrow (g\bar{\xi}, gSg^*; g\bar{\alpha}, g\Sigma g^*)$. Under this transformation, $(\Gamma, \Sigma) \longrightarrow (g^{*-1}\Gamma, g\Sigma g^*)$. It can be checked that these problems are invariant under the group of transformations G (with appropriate value of k). The principle of invariance in the complex case is exactly the same as it is in the real case. For the principle of invariance in the real case, the reader is referred to Lehman (1959). One can check that the maximal invariant under G in the space of $(\bar{\xi}, S)$ is $(R_1, \ldots, R_k)$ and the corresponding maximal invariant in the parametric case is $(\delta_1, \ldots, \delta_k)$.

A. The maximal invariant in the sample and parametric space under G (with k = 2) are (R_1, R_2) and (δ_1, δ_2) respectively. Under H_{10}: $\delta_2 = 0, \delta_1 > 0$ and under H_{11}: $\delta_1 > 0, \delta_2 > 0$. Proceeding exactly in the same way as in Giri (1964), pp. 188-189 one can show that the likelihood ratio test for this problem is uniformly most powerful invariant similar.

B. For this problem, the maximal invariant under G (with k = 3) in the sample space is (R_1, R_2, R_3) and the corresponding maximal invariant in the parametric space is $(\delta_1, \delta_2, \delta_3)$. Under H_{20}: $\delta_2 = \delta_3 = 0$, $\delta_1 > 0$ and H_{21}: $\delta_3 = 0$, $\delta_1 > 0$, $\delta_2 > 0$. Now proceeding exactly in the same way as in Giri (1965 a) pp. 1064-1065, we can conclude that the likelihood ratio test for Testing H_{20} against H_{21} is uniformly most powerful invariant similar.

C. This problem remains invariant under G (with K = 2). Under H_{30}: $\delta_2 = \delta_1 = 0$ and under H_{31}: $\delta_2 = 0$, $\delta_1 > 0$. The ratio of the density of (R_1, R_2) under H_{31} to its density under H_{30} is a monotonically increasing function of R_1 for $\delta_1 > 0$. Hence the likelihood ratio test in this case is uniformly most powerful invariant.

References

1 Giri, N. (1964), "On the likelihood ratio test of a normal multivariate testing problem", Ann.Math,Statist., 35, 181-189.

2 Giri, N. (1965), "On the likelihood ratio test of a normal multivariate testing problem II", Ann.Math.Statist.,36, 1061-1065.

3 Giri, N. (1965 a), "On the complex analogues of T^2 - and R^2 - tests", Ann.Math.Statist., 36, 664-670.

4 Lehmann, "E. L. (1959) Testing Statistical Hypothesis", Wiley, New York.

5 Salaevskii, O. V. (1968), "minimax character of Hotelling's T^2 test, Doki.Akad.Nawk.SSSR, 180, 733-735.

GLIVENKO-CANTELLI TYPE THEOREMS FOR DISTANCE FUNCTIONS BASED ON THE MODIFIED EMPIRICAL DISTRIBUTION FUNCTION OF M. KAC AND FOR THE EMPIRICAL PROCESS WITH RANDOM SAMPLE SIZE IN GENERAL[1]

Miklós Csörgő

McGill University

1. Summary and Introduction. In many probability models, one is concerned with a sequence $\{X_n : n \geq 1\}$ of independent random variables with a common distribution function, F say. When making statistical inferences within such a model, one frequently must do so on the basis of observations $X_1, X_2, \ldots X_N$ where the sample size N is a random variable. For example, N might be the number of observations that it was possible to take within a given period of time or within a fixed cost of experimentation. Much work has been done since 1951 on this problem for techniques based on the random sum of independent random variables $X_1 + X_2 + \ldots + X_N$: (See for example [3], [5], [6], [11], [13], [14], [7], [8] and their references). In [7], the independence condition is relaxed a little by considering random sums of random variables which are martingale differences and random sums of an arbitrary sequence of random variables are treated in [8]. Results have been also obtained for techniques based on max $(X_1, X_2, \ldots, X_n)$ by

[1]Work supported in part by the Canadian Mathematical Congress Summer Research Institute at McGill University and at the Universite de Montreal, Summer 1968.

Barndorff-Nielsen [4]. The asymptotic distribution, under random sample sizes, of statistics like those of Kolmogorov, Smirnov, Cramer and von Mises, which are distribution free statistics based upon the empirical distribution function, has been studied in [1], [2], [10], and [12]. In references [1], [2] and [10], the random sample sizes are Poisson random variables independent of the sample. In [12], $\{N\} = \{N_\lambda : \lambda \geq 0\}$ is a positive integer valued stochastic process satisfying

$$(1.1) \qquad N_\lambda / \lambda \rightarrow 1 \text{ in probability as } \lambda \rightarrow +\infty ,$$

and is not necessarily independent of $\{X_n : n \geq 1\}$. The same condition (1.1) is assumed in [3] and [13] too. The more general problem in which (1.1) is replaced by the condition : $N_\lambda / \lambda \rightarrow \gamma$ as $\lambda \rightarrow +\infty$ where γ is a strictly positive random variable is treated in [14], [11] [6], [7] and [8] for random partial sums of random variables.

It is the purpose of this paper to study the possibility of proving the Glivenko-Cantelli Theorem for distance functions based on the empirical distribution function with random sample size under the condition (1.1), which is the only condition used to prove convergence in distribution statements for this empirical process. It is proved in section 3 that the strong version of the Glivenko-Cantelli Theorem cannot be proved in general under the assumption of (1.1). Also in section 3, necessary and sufficient conditions are established for the behaviour of N_λ / λ and N_λ as $\lambda \rightarrow +\infty$ in terms of weak Glivenko-Cantelli type statements. The specific results of section 3 are based on some general laws of large numbers for random sequences of random variables which are established in section 2. In section 4, a simple proof of the original Glivenko-Cantelli Theorem ((3.2)) is given, which is based on the Hájek-Rényi inequality [9].

2. Some laws of large numbers for random sequences of random variables. The following two observations were proved in [7] and [8] respectively.

Proposition 1. Let $\{Z_n : n \geq 1\}$ be a sequence of random variables such that $Z_n \to 0$ with probability 1 as $n \to +\infty$ and let $\{N_\lambda : \lambda \geq 0$ be a positive integer valued stochastic process defined on the same probability space. If $N_\lambda \to +\infty$ with probability 1, then $Z_{N_\lambda} \to 0$ with probability 1 as $\lambda \to +\infty$.

Proposition 2. Let $\{Z_n : n \geq 1\}$ be a sequence of random variables such that $Z_n \to 0$ with probability 1 as $n \to \infty$ and let $\{N_\lambda : \lambda \geq 0\}$ be a positive integer valued stochastic process defined on the same probability space. If $N_\lambda \to +\infty$ in probability, then $Z_{N_\lambda} \to 0$ in probability as $\lambda \to +\infty$.

In [8] a counterexample is given which shows that if $Z_n \to 0$ in probability as $n \to +\infty$ and $N_\lambda \to +\infty$ in probability as $\lambda \to +\infty$, then Z_{N_λ} does not necessarily converge in probability to 0 as $\lambda \to +\infty$. Can it be true, however, in general that if $Z_n \to 0$ with probability 1 as $n \to +\infty$ and $N_\lambda \to +\infty$ in probability as $\lambda \to +\infty$, then $Z_{N_\lambda} \to 0$ with probability 1 as $\lambda \to +\infty$? The answer to this question is in the negative as will be seen from Theorem 1 and Corollary 1.

Theorem 1.

Let $\{Z_n : n \geq 1\}$ and $\{N_\lambda : \lambda \geq 0\}$ be as in Proposition 2. If $\lim_{\lambda \to +\infty} P\{N_\lambda < a\}$ exists for each $a > 0$, then for each $\varepsilon > 0$ and arbitrarily large positive a the following three statements are equivalent:

(2.1) $\lim_{\lambda \to +\infty} P\{N_\lambda < a\} = \lim_{\lambda \to +\infty} P\{|Z_{N_\lambda}| < \varepsilon, N_\lambda < a\}$,

(2.2) $\lim_{\lambda \to +\infty} P\{Z_{N_\lambda}| \geq \varepsilon, N_\lambda < a\} = 0$,

(2.3) $\lim_{\lambda \to +\infty} P\{|Z_{N_\lambda}| \geq \varepsilon\} = 0.$

Proof. For each λ, $\varepsilon > 0$ and $a > 0$ we have

(2.4) $P\{N_\lambda < a\} = P\{|Z_{N_\lambda}| < \varepsilon, N_\lambda < a\} + P\{|Z_{N_\lambda}| \geq \varepsilon, N_\lambda < a\},$

and the equivalence of statements (2.1) and (2.2) follows for each $\varepsilon > 0$ and $a > 0$. We consider now

(2.5) $P\{|Z_{N_\lambda}| \geq \varepsilon\} = P\{|Z_{N_\lambda}| \geq \varepsilon , N_\lambda < a\} + P\{|Z_{N_\lambda}| \geq \varepsilon , N_\lambda \geq a\}$.

If statement (2.3) is true then both terms of the right hand side of (2.5) must go to zero as $\lambda \to +\infty$ and so statement (2.2) must be also true. Conversely, if statement (2.2) is true then, to verify (2.3), we consider (2.5) and get from there

(2.6) $P\{|Z_{N_\lambda}| \geq \varepsilon\} = \Sigma_{n=a}^{\infty} P\{|Z_n| \geq \varepsilon , N_\lambda = n\} + P\{|Z_{N_\lambda}| \geq \varepsilon , N_\lambda < a\}$

$\leq P \sup_{a \leq n} |Z_n| \geq \varepsilon\} + P\{|Z_{N_\lambda}| \geq \varepsilon , N_\lambda < a\}$

Now given $\delta > 0$, arbitrarily small, we choose the value of $a > 0$ so large that the first probability statement of the last inequality becomes less than or equal to $\delta/2$. This can be done for $Z_n \to 0$ with probability 1 as $n \to +\infty$ is assumed in the statement of Theorem 1. Having picked this value of a, we choose the value of λ so large that the second probability statement of the inequality of (2.6) becomes less than or equal to $\delta/2$. This can be done for the statement of (2.2) is assumed to be true now for each $\varepsilon > 0$ and $a > 0$. This also completes the proof of Theorem 1.

Corollary 1. Let $\{Z_n : n \geq 1\}$ and $\{N_\lambda : \lambda \geq 0\}$ be as in Proposition 2. Then for each $\varepsilon > 0$ and $a > 0$ we have

$$\lim_{\lambda \to +\infty} P\{ N_\lambda < a\} = 0$$

if and only if

$$\lim_{\lambda \to +\infty} P\{|Z_{N_\lambda}| \geq \varepsilon\} = 0$$

and

$$\lim_{\lambda \to +\infty} P\{|Z_{N_\lambda}| < \varepsilon,\ N_\lambda < a\} = 0.$$

<u>Proof</u>. An obvious consequence of Theorem 1.

From Corollary 1 it follows now that the conclusion of Proposition 1 cannot be true in general under the condition that $N_\lambda \to +\infty$ in probability as $\lambda \to +\infty$, and so the negative answer to the above underlined question is confirmed.

Taking $Z_n = Y_n - Y$, where Y_n and Y are random variables, Corollary 1 says that if $Y_n \to Y$ with probability 1 as $n \to +\infty$ (and therefore, $Y_n \to Y$ in distribution as $n \to +\infty$) then $Y_{N_\lambda} \to Y$ in probability as $\lambda \to +\infty$(and, therefore, $Y_{N_\lambda} \to Y$ in distribution as $\lambda \to +\infty$) and

$$\lim_{\lambda \to +\infty} P\{|Y_{N_\lambda} - Y| < \varepsilon,\ N_\lambda < a\} = 0$$

if and only if $N_\lambda \to +\infty$ in probability as $\lambda \to +\infty$.

As immediate corollaries of Propositions 1 and 2 respectively we have the following two statements.

<u>Proposition 3</u>. Let $\{Z_n : n \geq 1$ and $\{N_\lambda : \lambda \geq 0\}$ be as in Proposition 1. If $N_\lambda/\lambda \to 1$ with probability 1, then $Z_{N_\lambda} \to 0$ with probability 1 as $\lambda \to +\infty$.

<u>Proposition 4</u>. Let $\{Z_n : n \geq 1\}$ and $\{N_\lambda : \lambda \geq 0\}$ be as in Proposition 2. If $N_\lambda/\lambda \to 1$ in probability as $\lambda \to +\infty$, then $Z_{N_\lambda} \to 0$ in probability as $\lambda \to +\infty$.

The counterexample given in [8] also shows that if $Z_n \to 0$ in probability as $n \to +\infty$ and $N_\lambda/\lambda \to 1$ in probability as $\lambda \to +\infty$, then Z_{N_λ} does not necessarily converge in probability to 0 as $\lambda \to +\infty$. Thus, even under the stronger assumption that $N_\lambda/\lambda \to 1$ in probability as $\lambda \to +\infty$(as compared to the requirement that $N_\lambda \to +\infty$ in probability as $\lambda \to +\infty$) we will have to require that $Z_n \to 0$ with probability 1 as $n \to +\infty$ in order to have $Z_{N_\lambda} \to 0$ in probability as

$\lambda \to +\infty$ (Proposition 4). Can it be true, however, that if $Z_n \to 0$ with probability 1 as $n \to +\infty$ and $N_\lambda/\lambda \to 1$ in probability as $\lambda \to +\infty$, then $Z_{N_\lambda} \to 0$ with probability 1 as $\lambda \to +\infty$? The answer to this question is also in the negative as will be seen from Theorem 2 and Corollary 2.

First we prove a weak law of large numbers type statement without restricting the behaviour of N_λ as $\lambda \to +\infty$.

Lemma 1. Let $\{Z_n : n \geq 1\}$ and $\{N_\lambda : \lambda \geq 0\}$ be as in Proposition 2. Then for each $\varepsilon > 0$

(2.7) $\lim_{\lambda \to +\infty} P\{|Z_{N_\lambda}| \geq \varepsilon, |N_\lambda/\lambda - 1| < \varepsilon\} = 0$

Proof. We have for each $\varepsilon > 0$

$$P\{|Z_{N_\lambda}| \geq \varepsilon, |N_\lambda/\lambda - 1| < \varepsilon\} = \sum_{|\frac{n}{\lambda} - 1| < \varepsilon} P\{|Z_n| \geq \varepsilon, N_\lambda = n\}$$

$$\leq P\{\sup_{[(1-\varepsilon)\lambda] + 1 \leq n} |Z_n| \geq \varepsilon\}$$

and, taking limits as $\lambda \to +\infty$ on both sides, (2.7) follows from the assumption that $Z_n \to 0$ with probability 1 as $n \to +\infty$.

Theorem 2.

Let $\{Z_n : n \geq 1\}$ and $\{N_\lambda : \lambda \geq 0\}$ be as in Proposition 2. If $\lim_{\lambda \to +\infty} P\{|N_\lambda/\lambda - 1| \geq \varepsilon\}$ exists for each $\varepsilon > 0$, then, for each $\varepsilon > 0$, the following three statements are equivalent:

(2.8) $\lim_{\lambda \to +\infty} P\{|N_\lambda/\lambda - 1| \geq \varepsilon\} = \lim_{\lambda \to +\infty} P\{|Z_{N_\lambda}| < \varepsilon, |N_\lambda/\lambda - 1| \geq \varepsilon\}$,

(2.9) $\lim_{\lambda \to +\infty} P\{|Z_{N_\lambda}| \geq \varepsilon, |N_\lambda/\lambda - 1| \geq \varepsilon\} = 0$,

(2.10) $\lim_{\lambda \to +\infty} P\{|Z_{N_\lambda}| \geq \varepsilon\} = 0$.

Proof. The proof of this theorem is similar to that of Theorem 1. Here one considers

(2.11) $P\{|N_\lambda/\lambda - 1| \geq \varepsilon\} = P\{|Z_{N_\lambda}| < \varepsilon, |N_\lambda/\lambda - 1| \geq \varepsilon\} + P\{|Z_{N_\lambda}| \geq \varepsilon, |N_\lambda/\lambda - 1| \geq \varepsilon\}$,

and the equivalence of statements (2.8) and (2.9) follows. The equivalence of (2.9) and (2.10) follows from Lemma 1 and

(2.12) $P\{|Z_{N_\lambda}| \geq \varepsilon\} = P\{|Z_{N\lambda}| \geq \varepsilon, |N_\lambda/\lambda - 1| < \varepsilon\} + P\{|Z_{N\lambda}| \geq \varepsilon, |N_\lambda/\lambda - 1| \geq \varepsilon\}$

Theorem 2 underlines the obvious fact that the converse of Proposition 4 is not necessarily true; that is to say if $Z_{N\lambda} \to 0$ in probability then N_λ/λ might or might not converge in probability to 1 as $\lambda \to +\infty$. If, however, N_λ/λ converges in probability to 1 as $\lambda \to +\infty$, then Theorem 2 implies the following interesting characterization of this assumption in this context.

<u>Corollary 2</u>. Let $\{Z_n : n \geq 1\}$ and $\{N_\lambda : \lambda \geq 0\}$ be as in Proposition 2. Then for each $\varepsilon > 0$

$$\lim_{\lambda \to +\infty} P\{|N_\lambda / \lambda - 1| \geq \varepsilon\} = 0$$

if and only if

$$\lim_{\lambda \to +\infty} P\{|Z_{N_\lambda}| \geq \varepsilon\} = 0$$

and

$$\lim_{\lambda \to +\infty} P\{|Z_{N_\lambda}| < \varepsilon, |N_\lambda/\lambda - 1| \geq \varepsilon\} = 0.$$

<u>Proof</u>. An obvious consequence of Theorem 2.

From Corollary 2 it follows now that the conclusion of Proposition 3 cannot be true in general under the condition that $N_\lambda/\lambda \to 1$ in probability as $\lambda \to +\infty$, and the negative answer to the second underlined question of this section is confirmed. Writing again $Z_n = Y_n - Y$, as we did immediately after Corollary 1, we get similar conclusions concerning convergence in distribution statements in this context.

We also remark that whenever the expression N_λ/λ occurs in this section, it can be replaced by $N_\lambda/f(\lambda)$, where $f(\lambda)$ is an arbitrary real valued function which increases monotonically to $+\infty$ as $\lambda \to +\infty$.

3. <u>On the Glivenko-Cantelli theorem under random sample sizes.</u> Let $\{X_n : n \geq 1\}$ be a sequence of independent random variables with a common distribution function F and let $\{N_\lambda : \lambda \geq 0\}$ be a positive integer valued stochastic process. Let $\psi_y(x)$ be 0 or 1 according as $x > y$ or $x \leq y$. Following M. Kac [10], we define

$$(3.1) \qquad F_\lambda^*(y) = \lambda^{-1} \sum_{j=1}^{N_\lambda} \psi_y(X_j), \qquad -\infty < y < +\infty,$$

where the sum is taken to be zero if $N_\lambda = 0$. In case N_λ is a Poisson random variable with mean value λ and independent of the sequence $\{X_n : n > 1\}$ then (3.1) is the modified empirical distribution function of M. Kac [10]. Let $F_n(y)$ be the ordinary empirical distribution function of the independent identically distributed random variables $X_1, \ldots, X_n$. In this context the Glivenko-Cantelli theorem says

$$(3.2) \qquad P\{\lim_{n \to +\infty} \sup_{-\infty < y < +\infty} |F_n(y) - F(y)| = 0\} = 1,$$

that is to say when the number of observations in a random sample increases ad infinitum, the ordinary empirical distribution function F_n converges uniformly over the whole real line to F with probability 1.

In applications when the number of observations N_λ is a random variable, which usually depends on $\{X_n : n \geq 1\}$, one would like to have a theorem which states the same thing as (3.2) under the condition

$$(3.3) \qquad N_\lambda / \lambda \to 1 \text{ in probability as } \lambda \to +\infty ;$$

that is to say, in analogy to the statement of (3.2), we would like to know if

$$(3.4) \qquad P\{\lim_{\lambda \to +\infty} \sup_{-\infty < y < +\infty} |F_\lambda^*(y) - F(y) = 0\} = 1$$

is true?

Before going into this problem we note that, defining the empirical distribution function of the random number of independent identically distributed random variables $X_1, X_2, \dots, X_N$ as

(3.5) $$F_{N_\lambda}(y) = N_\lambda^{-1} \sum_{j=1}^{N_\lambda} \psi_y(X_j), \qquad -\infty < y < +\infty ,$$

where this sum is taken to be zero if $N_\lambda = 0$ one has the obvious identity

(3.6) $$F_\lambda^* - F = F_{N_\lambda} - F + (N_\lambda/\lambda - 1) F_{N_\lambda} ,$$

which, as a consequence of (3.2) and Propositions 3 and 4, clearly converges uniformly over the whole real line in probability or with probability 1 depending on what N_λ/λ does as $\lambda \to +\infty$. Now the case of $N_\lambda/\lambda \to 1$ with probability 1 is not very interesting, for if one considers almost sure convergence then one ends up with non-random sequences of real functions (except on a set of measure zero) and such a strong assumption on N_λ/λ is not needed for convergence in distribution results. In the light of the results of [1], [2], [10], and [12] one would actually like to know if in general (3.4) is true under the condition (3.3), which is the only condition used to prove convergence in distribution statements for the empirical process with random sample size. The answer to this question is negative as will be seen from Theorem 3 which is fashioned after Corollary 2 of the previous section.

Theorem 3.

Let $\{X_n : n \geq 1\}$ be a sequence of independent random variables with a common distribution function F and let $\{N_\lambda : \lambda \geq 0\}$ be a positive integer valued stochastic process. Then for each $\varepsilon > 0$

$$(3.7) \qquad \lim_{\lambda \to +\infty} P\{|N_\lambda/\lambda - 1| \geq \varepsilon\} = 0$$

if and only if

$$\lim_{\lambda \to +\infty} P\{\sup_{-\infty < y < +\infty} |F_\lambda^*(y) - F(y)| > \varepsilon\} = 0$$

and

$$(3.8) \qquad \lim_{\lambda \to +\infty} P\{\sup_{-\infty<y<+\infty} |F_\lambda^*(y) - F(y)| < \varepsilon, |N_\lambda/\lambda - 1| \geq \varepsilon\} = 0.$$

Proof. We write

$$(3.9) \qquad P\{\sup_{-\infty<y<+\infty} |F_\lambda^*(y) - F(y)| \geq \varepsilon\}$$

$$= P\{\sup_{-\infty<y<+\infty} |F_\lambda^*(y) - F(y)| \geq \varepsilon, |N_\lambda/\lambda - 1| < \varepsilon\}$$

$$+ P\{\sup_{-\infty<y<+\infty} |F_\lambda^*(y) - F(y)| \geq \varepsilon, |N_\lambda/\lambda - 1| \geq \varepsilon\}.$$

The second term of this equality is bounded above by $P\{|N_\lambda/\lambda - 1| \geq \varepsilon\}$, which tends to zero as $\lambda \to +\infty$ if the condition (3.7) is assumed and we only have to show that the first term of the right-hand side of this equality also becomes arbitrarily small as $\lambda \to +\infty$. Concerning this term we get through the identity of (3.6)

$$(3.10) \qquad P\{\sup_{-\infty<y<+\infty} |F_\lambda^*(y) - F(y)| \geq \varepsilon, |N_\lambda/\lambda - 1| < \varepsilon\}$$

$$\leq P\{\sup_{-\infty<y<+\infty} |F_{N_\lambda}(y) - F(y)| + |N_\lambda/\lambda - 1| \geq \varepsilon, N_\lambda/\lambda - 1| < \varepsilon\}$$

$$\leq P\{\sup_{-\infty<y<+\infty} |F_{N_\lambda}(y) - F(y)| \geq \varepsilon/2, |N_\lambda/\lambda - 1| < \varepsilon\}$$

$$+ P\{|N_\lambda/\lambda - 1| \geq \varepsilon/2, |N_\lambda/\lambda - 1| < \varepsilon\}.$$

Now the second term of the last inequality of (3.10) is bounded above by $P\{|N_\lambda/\lambda - 1| \geq \varepsilon/2\}$, which goes to zero as $\lambda \to +\infty$, by (3.7) while the first term of this last inequality of (3.10) is bounded above by

$$(3.11) \qquad P\{\sup_{-\infty<y<+\infty} |F_{N_\lambda}(y) - F(y)| \geq \varepsilon/2\}$$

$$= P \sup_{-\infty < y < +\infty} |\sum_{j=1}^{N_\lambda} \psi_y(X_j) - N_\lambda F(y)| \quad /N_\lambda \geq \varepsilon /2\} \quad .$$

Writing $Z_{N_\lambda} = \sup_{-\infty < y < +\infty} | \sum_{j=1}^{N_\lambda} (X_j) - N_\lambda F(y) | /N_\lambda$ and using condition (3.7), it follows from Proposition 4 and (3.2) that $Z_{N_\lambda} \to 0$ in probability as $\lambda \to +\infty$.

We have, therefore, shown so far that condition (3.7) implies the first statement of (3.8) and that it also implies the second statement of (3.8) is obvious.

Conversely, if we assume now the two statements of (3.8) and write

$$(3.12) \quad P\{|N_\lambda/\lambda-1| \geq \varepsilon\} = P\{\sup_{-\infty < y < +\infty} |F_\lambda^*(y) - F(y)| \geq \varepsilon, |N_\lambda/\lambda-1| \geq \varepsilon\}$$

$$+ P\{\sup_{-\infty < y < +\infty} |F_\lambda^*(y) - F(y)| < \varepsilon, N\lambda/\lambda-1| \geq \varepsilon\} ,$$

then, assuming (3.8), it follows from (3.9) and (3.12) that the condition (3.7) is true, and this completes the proof of Theorem 3.

In the light of Theorem 3 we can thus conclude that the strong version of the Glivenko-Cantelli theorem (3.4) cannot be proved in general for the modified empirical distribution function of M. Kac under the condition (3.3), which is the only condition used to prove convergence in distribution statements for the empirical process with random sample size (see e.g. [1] , [2] , [10] and [12]).

We have already remarked that Theorem 3 was fashioned after Corollary 2 of Theorem 2, yet we proved Theorem 3 directly here. The reason for doing this was that Theorem 2 could not be straightforwardly applied to the randomized sequence $\{\sup_{-\infty < y < +\infty} |\lambda^{-1} \sum_{j=1}^{N} \psi_y(X_j) - F(y)|\}$ of Theorem 3, for it is not a completely randomized sequence of the ordinary empirical process

$\{\sup_{-\infty < y < +\infty} |n^{-1} \Sigma_{j=1}^{n} \psi_y (X_j) - F(y)|\}$ in the sense that, when randomizing, $F_n(y) = n^{-1} \Sigma_{j=1}^{n} \psi_y(X_j)$ is replaced by $F_\lambda^*(y)$ of (3.1) and not by $F_{N_\lambda}(y)$ of (3.5). Statements similar to Lemma 1 and Theorem 2 can, however, be proved here too. First we prove a weak law of large numbers type statement, similar to that of Lemma 1, without restricting the behaviour of N_λ as $\lambda \to +\infty$.

Lemma 2. Let $\{X_n : n \geq 1\}$ be a sequence of independent random variables with a common distribution function F and let $\{N_\lambda : \lambda \geq 0\}$ be a positive integer valued stochastic process. Then for each ε and δ, which are such that $\varepsilon > \delta > 0$, we have

$$(3.13) \quad \lim_{\lambda \to +\infty} P\{\sup_{-\infty < y < +\infty} |F_\lambda^*(y) - F(y)| \geq \varepsilon, |N_\lambda/\lambda - 1| < \delta\} = 0.$$

Proof. From the first inequality of (3.10) we get

$$(3.14) \quad P\{\sup_{-\infty < y < +\infty} |F_\lambda^*(y) - F(y)| \geq \varepsilon, |N_\lambda/\lambda - 1| < \delta\}$$

$$\leq \{P \sup_{-\infty < y < +\infty} |F_{N_\lambda}(y) - F(y)| \geq \varepsilon - \delta, |N_\lambda/\lambda - 1| < \delta\}$$

$$= \Sigma_{|\frac{n}{\lambda} - 1| < \delta} P\{\sup_{-\infty < y < +\infty} |F_n(y) - F(y)| \geq \varepsilon - \delta, N_\lambda = n\}$$

$$\leq P\{\sup_{[(1-\delta)\lambda]+1 \leq n} (\sup_{-\infty < y < +\infty} |F_n(y) - F(y)|) \geq \varepsilon - \delta\}$$

and, taking limits as $\lambda \to +\infty$ on both sides, (3.14) follows from the Glivenko Cantelli theorem of (3.2).

Using now Lemma 2, an argument similar to that of the proof of Theorem 2 gives

Corollary 3. Let $\{X_n : n \geq 1\}$ and $\{N_\lambda : \lambda \geq 0\}$ be as in Lemma 2. Then for each ε and δ, which are such that $\varepsilon > \delta > 0$, the following three statements are equivalent:

$$(3.15) \quad \lim_{\lambda \to +\infty} P\{|N_\lambda/\lambda - 1| \geq \delta\} = \lim_{\lambda \to +\infty} P\{\sup_{-\infty < y < +\infty} |F_\lambda^*(y) - F(y)| < \varepsilon,$$

$$|N_\lambda/\lambda - 1| \geq \delta\}$$

(3.16) $\lim_{\lambda \to +\infty} P\{\sup_{-\infty < y < +\infty} |F^*_\lambda(y) - F(y)| \geq \varepsilon\ ,\ |N_\lambda/\lambda - 1| \geq \delta\} = 0,$

(3.17) $\lim_{\lambda \to +\infty} P\{\sup_{-\infty < y < +\infty} |F^*_\lambda(y) - F(y)| \geq \varepsilon\} = 0.$

As a result of Corollary 3, we can again prove Theorem 3 from this statement.

Similar Glivenko-Cantelli type statements can be proved for the empirical process with random sample size based on the sample distribution function of (3.5). In fact in this case we have immediately available results as corollaries of the propositions of section 2. Let

(3.18) $\qquad Z_n = \sup_{-\infty < y < +\infty} | F_n(y) - F(y)|$

and

(3.19) $\qquad Z_{N_\lambda} = \sup_{-\infty < y < +\infty} |F_{N_\lambda}(y) - F(y)|,$

where $F_n(y) = n^{-1} \sum_{j=1}^{n} \psi_y(X_j)$ and $F_{N_\lambda}(y)$ is as in (3.5), both of them based on a sequence $\{X_n : n \geq 1\}$ of independent identically distributed random variables. Then Propositions 1, 2, 3 and 4, Theorems 1 and 2, Corollaries 1 and 2 and Lemma 1 section 2 hold; for, in this case Z_{N_λ} of (3.19) is the completely randomized version of the Z_n sequence of (3.18) which satisfies the strong law of large numbers in the sense of the Glivenko-Cantelli theorem of (3.2). Again the main conclusion is that the strong version of the Glivenko-Cantelli theorem cannot be proved in general for the empirical process with random sample size of (3.19) under the condition (3.3), which is the only condition used to prove convergence in distribution statements for this process (see e.g. [12]).

4. Proof of the Glivenko-Cantelli theorem ((3.2)) using the Hájek-Rényi Inequality. The statement of (3.2) is equivalent to

(4.1) $\lim_{n \to +\infty} P\{\sup_{m \geq n} (\sup_{-\infty < y < +\infty} |F_m(y) - F(y)|) \geq \varepsilon\} = 0$

for each $\varepsilon > 0$. Let M be an arbitrarily large natural number and let $y_{M,k}$, $k = 1, 2, \ldots, M$, be the smallest number y for which we have $F(y - 0) < \frac{k}{M} \leq F(y)$. Then

(4.2) $$\sup_{-\infty < y < +\infty} |F_n(y) - F(y)| \leq \max(\Delta_n^{(1)}, \Delta_n^{(2)}) + \frac{1}{M},$$

where

$$\Delta_n^{(1)} = \max_{1 \leq k \leq M} |F_n(y_{M,k}) - F(y_{M,k})|$$

and

$$\Delta_n^{(2)} = \max_{1 \leq k \leq M} |F_n(y_{M,k} - 0) - F(y_{M,k} - 0)|.$$

On choosing the value of M so large that $\frac{1}{M} < \varepsilon$ and using the inequality of (4.2) we get

$$P\{\sup_{m \geq n} (\sup_{-\infty < y < +\infty} |F_m(y) - F(y)|) \geq \varepsilon\}$$

$$\leq P\{\sup_{m \geq n} (\max(\Delta_m^{(1)}, \Delta_m^{(2)}) + \frac{1}{M}) \geq \varepsilon\}$$

$$\leq P\{\sup_{m \geq n} \Delta_m^{(1)} \geq \varepsilon - \frac{1}{M}\} + P\{\sup_{m \geq n} \Delta_m^{(2)} \geq \varepsilon - \frac{1}{M}\}$$

$$= \sum_{k=1}^{M} P\{\sup_{m \geq n} |F_m(y_{M,k}) - F(y_{M,k})| \geq \varepsilon - \frac{1}{M}\}$$

$$+ \sum_{k=1}^{M} P\{\sup_{m \geq n} |F_m(y_{M,k} - 0) - F(y_{M,k} - 0)| \geq \varepsilon - \frac{1}{M}\}$$

$$\leq \frac{2M}{\left(\varepsilon - \frac{1}{M}\right)^2} \left\{ \frac{1}{n} + \sum_{m=n+1}^{\infty} \frac{1}{m^2} \right\},$$

where the last inequality is the result of applying the Hájek-Rényi inequality [9] to $m^{-1} \mid \sum_{j=1}^{m} (\psi_{y_{M,k}}(X_j) - F(y_{M,k}))\mid$ and

$m^{-1} \mid \sum_{j=1}^{m} (\psi_{y_{M,k}-0}(X_j) - F(y_{M,k}-0))$ respectively M times.

Taking limits as $n \to +\infty$, the statement of (3.2) follows.

References

1 Allen, J. L. and Beekman, J. A. (1966). A statistical test involving a random variables. Ann. Math. Statist. 37 1305-1309.

2 Allen, J. L. and Beekman, J. A. (1967). Distribution of a M. Kac Statistic. Ann. Math. Statist. 38 1919-1923.

3 Anscombe, F. J. (1952). Large-sample theory of sequential estimation. Proc. Cambridge Philos. Soc. 48 600-607.

4 Barndorff-Nielsen, O. (1964). On the limit distribution of the maximum of a random number of independent random variables. Acta Math. Acad. Sci. Hung. 15, 399-403.

5 Billingsley, Patrick (1962). Limit theorems for randomly selected partial sums. Ann. Math. Statist. 33 85-92.

6 Blum, J. R., Hanson, D. L. and Rosenblatt, J. I (1963). On the central limit theorem for the sum of a random number of independent random variables. Z. Wahrscheinlichkeitstheorie und Verw. Gebiete, 1 389-393.

7 Csörgő, Miklós (1968). On the strong law of large numbers and the central limit theorem for martingales. Trans. Amer. Math. Soc. 131 259-275.

8 Csorgo, Miklos and Fischler, Roger (1967). Departure from independence; the strong law, standard and random-sum central limit theorems. (Unpublished)

9 Hájek, J. and Rényi, A. (1955). Generalization of an inequality of Kolmogorov. Acta Math. Acad. Sci. Hung. 6 281-283.

10 Kac, M. (1949). On deviations between theoretical and empirical distributions. Proc. Nat. Acad. Sci. U.S.A. 35 252-257.

11 Mogyoródi, J. (1962). A central limit theorem for the sum of a random number of independent random variables. Publications of the Math. Inst. Hung. Acad. Sci., Series A 7 409-424.

12 Pyke, Ronald (1966). The weak convergence of the empirical process with random sample size. (Unpublished).

13 Rényi, A. (1957). On the asumptotic distribution of the sum of a random number of independent random variables. Acta Math. Acad. Sci. Hung. 8 193-197.

14 Rényi, A. (1960). On the central limit theorem for the sum of a random number of independent random variables. Acta Math. Acad. Sci. Hung. 11 97-102.

A Complete Metric Space of Sub-σ-Algebras

J.M. Singh
McMaster University

Summary: In this paper it is shown that a metric can be defined on the set of all sub-σ-algebras of a given σ-algebra. It is observed that C. Rajski's Theorem ([9]) on the metric space of discrete probability distributions turns out to be a particular case of our theorem which also provides a much shorter proof. The completeness and other properties of this metric space are also established.

§ 1 Conditional Entropies of Sub-σ-Algebras

1.1 Introduction

Throughout the discussion of this paper, unless otherwise stated, the probability space under consideration is denoted by $(\Omega, \mathcal{R}, P)$ where Ω is the abstract set, $\mathcal{R}$ is the σ-algebra of all subsets of Ω and P is the probability measure over $\mathcal{R}$. By a finite σ-algebra we shall mean a σ-algebra consisting of a finite number of subsets of Ω. It is easy to verify that there is a one-to-one correspondence between finite measurable partitions of Ω and finite sub-σ-algebras of $\mathcal{R}$ (cf. [7], P. 7).

Notation 1.1.1: If $\{\mathcal{R}_\alpha : \alpha \in \Lambda\}$ is a family of sub-σ-algebras of $\mathcal{R}$, then, $\bigvee_{\alpha \in \Lambda} \mathcal{R}_\alpha$ denotes the sub-σ-algebra of $\mathcal{R}$ generated by $\bigcup_{\alpha \in \Lambda} \mathcal{R}_\alpha$ and $\bigwedge_{\alpha \in \Lambda} \mathcal{R}_\alpha$ denotes the largest sub-σ-algebra of $\mathcal{R}$ contained in each of the sub-σ-algebra $\mathcal{R}_\alpha$.

Definition 1.1.1: Let $\mathcal{R}_0 \subseteq \mathcal{R}$ be the finite σ-algebra whose atoms are $A_1, A_2, \ldots, A_r$. The entropy $H(\mathcal{R}_0)$ of the finite σ-algebra $\mathcal{R}_0$ is defined as

$$H(\mathcal{R}_0) = - \sum_{k=1}^{r} p_k \log_2 p_k, \quad \text{where } p_k = P(A_k); \quad k = 1,2,\ldots, r.$$

Definition 1.1.2: Let $\mathcal{R}_0$, $\mathcal{R}_0'$ be two finite sub-σ-algebras of $\mathcal{R}$ whose atoms are $A_i (1 \leq i \leq r)$ and $A_k' (1 \leq k \leq r')$ respectively. We define

$$p_{ik} = \begin{cases} \dfrac{P(A_i \cap A_k')}{P(A_i)} & \text{if } P(A_i) > 0 \\ \dfrac{1}{r'} \text{ (say)} & \text{if } P(A_i) = 0 \end{cases}$$

and $P^{(i)} = (p_{i1}, p_{i2}, \ldots, p_{ir'})$.

The conditional entropy $H(\mathcal{R}_0'/\mathcal{R}_0)$ of the finite sub-σ-algebra $\mathcal{R}_0'$ w.r.t. the finite sub-σ-algebra $\mathcal{R}_0$ is defined as

$$H(\mathcal{R}_0'/\mathcal{R}_0) = \sum_{i=1}^{r} p_i \, H(P^{(i)}), \text{ where } H(P^{(i)}) = - \sum_{k=1}^{r'} p_{ik} \log_2 p_{ik}$$

Definition 1.1.3: Let S be the system of all finite sub-σ-algebras of any given σ-algebra $\mathcal{R}_0 \subseteq \mathcal{R}$. The entropy $H(\mathcal{R}_0)$ of the σ-algebra $\mathcal{R}_0$ is defined as

$$H(\mathcal{R}_0) = \sup_{C \in S} H(C)$$

Definition 1.1.4: Let S and S' be the systems of all finite sub-σ-algebras of $\mathcal{R}_0 \subseteq \mathcal{R}$ and $\mathcal{R}_0' \subseteq \mathcal{R}$ respectively; then the conditional entropy $H(\mathcal{R}_0'/\mathcal{R}_0)$ of $\mathcal{R}_0'$ w.r.t $\mathcal{R}_0$ is defined as

$$H(\mathcal{R}_0'/\mathcal{R}_0) = \sup_{C' \in S'} \inf_{C \in S} H(C'/C)$$

Remark 1.1.1: Let $\mathcal{R}_0$, $\mathcal{R}_1$, $\mathcal{R}_0'$, $\mathcal{R}_1'$ be arbitrary σ-algebras $\subseteq \mathcal{R}$; then we have the following (cf. [5], P. 260, Theorem 4).

A. Equalities

i) $H(\mathcal{R}_0 \vee \mathcal{R}_0') = H(\mathcal{R}_0) + H(\mathcal{R}_0'/\mathcal{R}_0)$.

ii) $H(\mathcal{R}_0' \vee \mathcal{R}_1' / \mathcal{R}_0) = H(\mathcal{R}_0'/\mathcal{R}_0) + H(\mathcal{R}_1'/\mathcal{R}_0' \vee \mathcal{R}_0)$.

iii) $H(\mathcal{R}_0'/\mathcal{R}_0) = H(\mathcal{R}_0' \vee \mathcal{R}_1'/\mathcal{R}_0)$ if $\mathcal{R}_1' \subseteq \mathcal{R}_0$.

iv) $H(\mathcal{R}_0'/\mathcal{R}_0) = 0$ if $\mathcal{R}_0' \subseteq \mathcal{R}_0$ up to equivalence.

B. Inequalities

v) $H(\mathcal{R}_0) \leq H(\mathcal{R}_1)$ if $\mathcal{R}_0 \subset \mathcal{R}_1$.

vi) $H(\mathcal{R}_0'/\mathcal{R}_0) \leq H(\mathcal{R}_1'/\mathcal{R}_0)$ if $\mathcal{R}_0' \subseteq \mathcal{R}_1'$.

vii) $H(\mathcal{R}_0'/\mathcal{R}_0) \geq H(\mathcal{R}_0'/\mathcal{R}_1)$ if $\mathcal{R}_0 \subseteq \mathcal{R}_1$.

viii) $H(\mathcal{R}_0' \vee \mathcal{R}_1'/\mathcal{R}_0) \leq H(\mathcal{R}_0'/\mathcal{R}_0) + H(\mathcal{R}_1'/\mathcal{R}_0)$.

Remark 1.1.2: If Z is the set of equivalence classes of sub-σ-algebras of $\mathcal{R}$ with finite entropy, then (Z,d) is a metric space where d is given by

$$d(\mathcal{R}_0, \mathcal{R}_0') = H(\mathcal{R}_0/\mathcal{R}_0') + H(\mathcal{R}_0'/\mathcal{R}_0); \quad \mathcal{R}_0, \mathcal{R}_0' \in Z.$$

(cf. [5] , Theorem 7, P. 265).

1.2 Generalized Conditional Entropies

A few generalized conditional entropies are given by M. Behara and P. Nath for finite measureable partitions (See [1]) from which Renyi's conditional entropy and Shannon's conditional entropy can be deduced. Here, we define one of these generalized conditional entropies.

If $\mathcal{R}_0'$ and $\mathcal{R}_0$ are two finite sub-σ-algebras of $\mathcal{R}$ whose atoms are $A_k'(1 \leq k \leq r')$ and $A_i(1 \leq i \leq r)$, then we know that the conditional probability of A given $\mathcal{R}_0$ is defined up to equivalence by

$$P_{\mathcal{R}_0}(A) = P(A/\mathcal{R}_0) = \sum_{i=1}^{r} \frac{P(A \cap A_i)}{P(A_i)} \chi(A_i); \quad A \in \mathcal{R}_0'.$$

where $\chi(A_i)$ is the characteristic function of A_i and the conditional entropy of $\mathcal{R}_0'$ w.r.t $\mathcal{R}_0$ is defined as

$$I_\alpha(\mathcal{R}_0'/\mathcal{R}_0) = \sum_{k=1}^{r'} \int_\Omega P^{\alpha-1}(\mathcal{R}_0)\, Z_\alpha(P(A_k'/\mathcal{R}_0))\, dP.$$

$$\text{where } z_\alpha(t) = \begin{cases} \dfrac{t-t^\alpha}{1-2^{1-\alpha}} \; ; & t \quad (0,1], \quad \alpha\in[0,\infty) \\ 1 \; ; & t = 0, \alpha = 0 \\ 0 \; ; & t = 0, \alpha \in (0,\infty). \end{cases}$$

If $\mathcal{R}_0, \mathcal{R}_0', \mathcal{R}_0''$ be the finite sub-σ-algebras of $\mathcal{R}$, then the following properties hold; (cf. [1], P. 34-37).

i) $\mathcal{R}_0' \subseteq \mathcal{R}_0$ up to eqivalence $\Longleftrightarrow$ $I_\alpha(\mathcal{R}_0'/\mathcal{R}_0) = 0$; $\alpha > 0$.

ii) $\mathcal{R}_0' \subseteq \mathcal{R}_0''$ $\Rightarrow$ $I_\alpha(\mathcal{R}_0/\mathcal{R}_0'') \leq I_\alpha(\mathcal{R}_0/\mathcal{R}_0')$; $\alpha \geq 1$.

iii) $I_\alpha(\mathcal{R}_0' \vee \mathcal{R}_0''/\mathcal{R}_0) = I_\alpha(\mathcal{R}_0'/\mathcal{R}_0) + I_\alpha(\mathcal{R}_0''/\mathcal{R}_0 \vee \mathcal{R}_0')$; $\alpha \geq 0$.

iv) $I_\alpha(\mathcal{R}_0 \vee \mathcal{R}_0') = I_\alpha(\mathcal{R}_0) + I_\alpha(\mathcal{R}_0'/\mathcal{R}_0)$; $\alpha \geq 0$.

v) If $\mathcal{R}_0 \subseteq \mathcal{R}_0'$, then $I_\alpha(\mathcal{R}_0'/\mathcal{R}_0) = I_\alpha(\mathcal{R}_0') - I_\alpha(\mathcal{R}_0)$; $\alpha \geq 0$.

vi) If $\mathcal{R}_0' \subseteq \mathcal{R}_0''$, then $I_\alpha(\mathcal{R}_0'/\mathcal{R}_0) \leq I_\alpha(\mathcal{R}_0''/\mathcal{R}_0)$; $\alpha \geq 0$.

A few more generalized conditional entropies are also discussed in [1].

1.3 An Integral Representation of Conditional Entropy

Definition 1.3.1: Let $\mathcal{B}$ and $\mathcal{C}$ be finite sub-σ-algebras of $\mathcal{R}$ generated by the measureable partitions $\{B_i\}_{i=1}^n$ and $\{C_j\}_{j=1}^m$ respectively. Let $\mathcal{R}_0$ be a sub-σ-algebra of $\mathcal{R}$. Then the function $H_{\mathcal{R}_0}(\mathcal{B}/\mathcal{C})$ is almost everywhere defined by

$$H_{\mathcal{R}_0}(\mathcal{B}/\mathcal{C})(x) = \sum_{j=1}^{m} P_{\mathcal{R}_0}(C_j)(x) \sum_{i=1}^{n} z\left(\frac{P_{\mathcal{R}_0}(B_i \cap C_j)(x)}{P_{\mathcal{R}_0}(C_j)(x)} \right)$$

where $z(x) = -x \log_2 x$, $0 < x \leq 1$

and $z(0) = 0$.

In the remark below we shall show that the conditional entropy can be represented as the integral of the function $H_{\mathcal{R}_0}(\mathcal{B}/\mathcal{C})(x)$.

<u>Remark 1.3.1:</u> The following properties hold; (cf. [8] , Proposition 2.1)

a) $H_{\mathcal{R}_0}(\mathcal{B}/\mathcal{C})$ is a $\mathcal{R}_0$-measureable function on Ω .

b) For almost every $x \in \Omega$ we conclude by the definition of $P_{\mathcal{R}_0}$ (cf. [2] , Chapter II, P. 95) that the mapping $P_{\mathcal{R}_0} : A \longrightarrow P_{\mathcal{R}_0}(A)(x)$ is a measure on $\mathcal{B}_1 \vee \mathcal{C}_1$ and it follows that for any pair of finite sub-σ-algebras $\mathcal{B}_1$ and $\mathcal{C}_1$ of $\mathcal{R}$; $\mathcal{B}_1 \supset \mathcal{B}$ and $\mathcal{C}_1 \supset \mathcal{C}$, $H_{\mathcal{R}_0}(\mathcal{B}/\mathcal{C}_1)$, $H_{\mathcal{R}_0}(\mathcal{B}_1/\mathcal{C}_1)$ and $H_{\mathcal{R}_0}(\mathcal{B}_1/\mathcal{C})$ are conditional entropies w.r.t the measure space $(\Omega, \mathcal{B}_1 \vee \mathcal{C}_1, P_{\mathcal{R}_0})$ and the following result is true

$$H_{\mathcal{R}_0}(\mathcal{B}/\mathcal{C}_1) \le H_{\mathcal{R}_0}(\mathcal{B}_1/\mathcal{C}_1) \le H_{\mathcal{R}_0}(\mathcal{B}_1/\mathcal{C}) \text{ a.e}$$

c) $H(\mathcal{B}/\mathcal{R}_0 \vee \mathcal{C}) = \int_\Omega H_{\mathcal{R}_0}(\mathcal{B}/\mathcal{C})\, dP.$

<u>Remark 1.3.2:</u> If $\mathcal{B}$, $\mathcal{C}$, $\mathcal{B}_1$ and $\mathcal{C}_1$ are finite sub-σ-algebras of $\mathcal{R}$, then we have the following results (cf. [12] , Theorem 2.2.2, P. 17)

1. $H_{\mathcal{R}_0}(\mathcal{B} \vee \mathcal{C}) = H_{\mathcal{R}_0}(\mathcal{B}) + H_{\mathcal{R}_0}(\mathcal{C}/\mathcal{B})$ a.e.
2. $H_{\mathcal{R}_0}(\mathcal{B} \vee \mathcal{C}/\mathcal{B}_1) = H_{\mathcal{R}_0}(\mathcal{B}/\mathcal{B}_1) + H_{\mathcal{R}_0}(\mathcal{C}/\mathcal{B} \vee \mathcal{B}_1)$ a.e.
3. $H_{\mathcal{R}_0}(\mathcal{B}/\mathcal{C}) = H_{\mathcal{R}_0}(\mathcal{B} \vee \mathcal{B}_1/\mathcal{C})$ a.e if $\mathcal{B}_1 \subseteq \mathcal{C}$.
4. $H_{\mathcal{R}_0}(\mathcal{B}/\mathcal{C}) = 0$ a.e. if $\mathcal{B} \subseteq \mathcal{C}$ up to equivalence.
5. $H_{\mathcal{R}_0}(\mathcal{B}) \le H_{\mathcal{R}_0}(\mathcal{C})$ a.e. if $\mathcal{B} \subseteq \mathcal{C}$.
6. $H_{\mathcal{R}_0}(\mathcal{B}/\mathcal{C}) \le H_{\mathcal{R}_0}(\mathcal{B}_1/\mathcal{C})$ a.e. if $\mathcal{B} \subseteq \mathcal{B}_1$
7. $H_{\mathcal{R}_0}(\mathcal{B}/\mathcal{C}) \ge H_{\mathcal{R}_0}(\mathcal{B}/\mathcal{C}_1)$ a.e. if $\mathcal{C} \subseteq \mathcal{C}_1$.

§2 A Complete Metric Space of Sub-σ-Algebras

2.1 Metric Space of Sub-σ-Algebras

C. Rajski proved that the functional

(1) $d(x,y) = 1 - \frac{I(x,y)}{H(x,y)}$ where $H(x,y) \neq 0$ and $I(x,y) = H(x) + H(y) - H(x,y)$

is a distance in the set X of all discrete probability distributions (cf. [9] , Theorem P. 372). It is a consequence of this theorem, that in Information Theory the dependence between the transmitted and the received discrete signals may be expressed as a distance. Replacing x by $\mathcal{R}_0 \subseteq \mathcal{R}$ and y by $\mathcal{R}_0' \subseteq \mathcal{R}$ in (1), we prove in Theorem 2.1.1 that,

$$d(\mathcal{R}_0, \mathcal{R}_0') = 1 - \frac{I(\mathcal{R}_0, \mathcal{R}_0')}{H(\mathcal{R}_0 \vee \mathcal{R}_0')}, \; H(\mathcal{R}_0 \vee \mathcal{R}_0') \neq 0$$

is a metric in the set of all equivalence classes of sub-σ-algebras of $\mathcal{R}$.

It is observed that the theorem given by C. Rajski ([9]) is a particular case of Theorem 2.1.1 and the proof is by comparison concise. In order to show this, we need the following lemma:

Lemma 2.1.1: If $(\Omega, \mathcal{F}, \mu)$ is a probability space where $\Omega = \{w: 0 \leq w \leq 1\}$, $\mathcal{F}$ is the σ-algebra consisting of all Borel subsets of Ω and μ is Lebesgue measure, then corresponding to every discrete probability distribution there exists a sub-σ-algebra of $\mathcal{F}$.

Proof: A discrete probability distribution is the collection of various values of a random variable which correspond to the atoms of a finite or countable measurable partition as the case may be together with the probability measure of these atoms.

To Prove the lemma, we consider the following discrete probability distribution:

$$\text{Prob}(X = x_i) = p_i;\ i = 1,2,\ldots;\ \sum_i p_i = 1.$$

Let the atoms corresponding to the values $x_1, x_2, x_3, \ldots$ of random variables be as follows:

$$\{x:\ 0 \le x < p_1\},\ \{x:\ p_1 \le x < p_1+p_2\},\ \{x:\ p_1+p_2 \le x < p_1+p_2+p_3\},\ \ldots$$

The above subsets of Ω, obviously forms a finite or countable measurable partition of $\Omega = [0,1]$ and hence there exists a sub-σ-algebra of $\mathcal{F}$ corresponding to this countable measurable partition.

<u>Theorem 2.1.1</u>: If Z is the set of equivalence classes of sub-σ-algebras of $\mathcal{R}$ with finite entropy then the functional

$$d(\mathcal{R}_0, \mathcal{R}_0') = 1 - \frac{I(\mathcal{R}_0, \mathcal{R}_0')}{H(\mathcal{R}_0 \vee \mathcal{R}_0')}\ ;\quad H(\mathcal{R}_0 \vee \mathcal{R}_0') \ne 0$$

is a distance in the set Z.

<u>Proof</u>:

a) $$d(\mathcal{R}_0, \mathcal{R}_0') = 1 - \frac{I(\mathcal{R}_0, \mathcal{R}_0')}{H(\mathcal{R}_0 \vee \mathcal{R}_0')} = 1 - \frac{H(\mathcal{R}_0) + H(\mathcal{R}_0') - H(\mathcal{R}_0 \vee \mathcal{R}_0')}{H(\mathcal{R}_0 \vee \mathcal{R}_0')}$$

$$= \frac{\{H(\mathcal{R}_0 \vee \mathcal{R}_0') - H(\mathcal{R}_0')\} + \{H(\mathcal{R}_0 \vee \mathcal{R}_0') - H(\mathcal{R}_0)\}}{H(\mathcal{R}_0 \vee \mathcal{R}_0')}$$

$$= \frac{H(\mathcal{R}_0/\mathcal{R}_0') + H(\mathcal{R}_0'/\mathcal{R}_0)}{H(\mathcal{R}_0 \vee \mathcal{R}_0')} \geqslant 0.$$

b) $$d(\mathcal{R}_0, \mathcal{R}_0') = 0 \Rightarrow H(\mathcal{R}_0'/\mathcal{R}_0) + H(\mathcal{R}_0/\mathcal{R}_0') = 0$$

$\Rightarrow \mathcal{R}_0' \subseteq \mathcal{R}_0$ up to equivalence and

$\mathcal{R}_0 \subseteq \mathcal{R}_0'$ up to equivalence

$\Rightarrow \mathcal{R}_0 = \mathcal{R}_0'$ up to equivalence.

c) Let $\mathcal{R}_0 = \mathcal{R}_0'$ up to equivalence.

therefore $H(\mathcal{R}_0/\mathcal{R}_0') = 0$ and $H(\mathcal{R}_0'/\mathcal{R}_0) = 0$

$$\Rightarrow \frac{H(\mathcal{R}_0/\mathcal{R}_0') + H(\mathcal{R}_0'/\mathcal{R}_0)}{H(\mathcal{R}_0 \vee \mathcal{R}_0')} = 0 \Rightarrow d(\mathcal{R}_0, \mathcal{R}_0') = 0.$$

d) Now we establish the triangle inequality. We have

$$H(\mathcal{R}_0 \vee \mathcal{R}_0') \leq H(\mathcal{R}_0 \vee \mathcal{R}_0' \vee \mathcal{R}_0''); \quad \mathcal{R}_0, \mathcal{R}_0', \mathcal{R}_0'' \in Z$$

$$\Rightarrow \frac{H(\mathcal{R}_0 \vee \mathcal{R}_0') - H(\mathcal{R}_0')}{H(\mathcal{R}_0 \vee \mathcal{R}_0')} \leq \frac{H(\mathcal{R}_0 \vee \mathcal{R}_0' \vee \mathcal{R}_0'') - H(\mathcal{R}_0')}{H(\mathcal{R}_0 \vee \mathcal{R}_0' \vee \mathcal{R}_0'')}$$

$$\Rightarrow \frac{H(\mathcal{R}_0/\mathcal{R}_0')}{H(\mathcal{R}_0 \vee \mathcal{R}_0')} \leq \frac{H(\mathcal{R}_0 \vee \mathcal{R}_0' \vee \mathcal{R}_0'') - H(\mathcal{R}_0')}{H(\mathcal{R}_0 \vee \mathcal{R}_0' \vee \mathcal{R}_0'')}$$

$$= \frac{H(\mathcal{R}_0 \vee \mathcal{R}_0' \vee \mathcal{R}_0'') - H(\mathcal{R}_0' \vee \mathcal{R}_0'') + H(\mathcal{R}_0' \vee \mathcal{R}_0'') - H(\mathcal{R}_0')}{H(\mathcal{R}_0 \vee \mathcal{R}_0' \vee \mathcal{R}_0'')}$$

$$= \frac{H(\mathcal{R}_0/\mathcal{R}_0' \vee \mathcal{R}_0'')}{H(\mathcal{R}_0 \vee \mathcal{R}_0' \vee \mathcal{R}_0'')} + \frac{H(\mathcal{R}_0''/\mathcal{R}_0')}{H(\mathcal{R}_0 \vee \mathcal{R}_0' \vee \mathcal{R}_0'')} \leq \frac{H(\mathcal{R}_0/\mathcal{R}_0' \vee \mathcal{R}_0'')}{H(\mathcal{R}_0 \vee \mathcal{R}_0' \vee \mathcal{R}_0'')} + \frac{H(\mathcal{R}_0''/\mathcal{R}_0')}{H(\mathcal{R}_0' \vee \mathcal{R}_0'')}$$

Hence $\dfrac{H(\mathcal{R}_0/\mathcal{R}_0')}{H(\mathcal{R}_0 \vee \mathcal{R}_0')} \leq \dfrac{H(\mathcal{R}_0/\mathcal{R}_0' \vee \mathcal{R}_0'')}{H(\mathcal{R}_0 \vee \mathcal{R}_0' \vee \mathcal{R}_0'')} + \dfrac{H(\mathcal{R}_0''/\mathcal{R}_0')}{H(\mathcal{R}_0' \vee \mathcal{R}_0'')}$. ——— (1)

Now, we prove that $\dfrac{H(\mathcal{R}_0/\mathcal{R}_0' \vee \mathcal{R}_0'')}{H(\mathcal{R}_0 \vee \mathcal{R}_0' \vee \mathcal{R}_0'')} \leq \dfrac{H(\mathcal{R}_0/\mathcal{R}_0'')}{H(\mathcal{R}_0 \vee \mathcal{R}_0'')}$. ———(2)

We have $H(\mathcal{R}_0 \vee \mathcal{R}_0' \vee \mathcal{R}_0'') = H(\mathcal{R}_0 \vee \mathcal{R}_0'') + H(\mathcal{R}_0'/\mathcal{R}_0 \vee \mathcal{R}_0'')$

$$\Rightarrow H(\mathcal{R}_0 \vee \mathcal{R}_0' \vee \mathcal{R}_0'') \geq H(\mathcal{R}_0 \vee \mathcal{R}_0'') \Rightarrow \frac{H(\mathcal{R}_0/\mathcal{R}_0' \vee \mathcal{R}_0'')}{H(\mathcal{R}_0 \vee \mathcal{R}_0' \vee \mathcal{R}_0'')} \leq \frac{H(\mathcal{R}_0/\mathcal{R}_0' \vee \mathcal{R}_0'')}{H(\mathcal{R}_0 \vee \mathcal{R}_0'')}$$

$$\Rightarrow \frac{H(\mathcal{R}_0/\mathcal{R}_0' \vee \mathcal{R}_0'')}{H(\mathcal{R}_0 \vee \mathcal{R}_0' \vee \mathcal{R}_0'')} \leq \frac{H(\mathcal{R}_0/\mathcal{R}_0'')}{H(\mathcal{R}_0 \vee \mathcal{R}_0'')}.$$

Thus (2) is proved.

Now from (1) and (2) we have

$$\frac{H(\mathcal{R}_0/\mathcal{R}_0')}{H(\mathcal{R}_0 V \mathcal{R}_0')} \leq \frac{H(\mathcal{R}_0/\mathcal{R}_0'')}{H(\mathcal{R}_0 V \mathcal{R}_0'')} + \frac{H(\mathcal{R}_0''/\mathcal{R}_0')}{H(\mathcal{R}_0' V \mathcal{R}_0'')} . \text{ ——— } (3)$$

Interchanging the roles of $\mathcal{R}_0$ and $\mathcal{R}_0'$, we obtain

$$\frac{H(\mathcal{R}_0'/\mathcal{R}_0)}{H(\mathcal{R}_0 V \mathcal{R}_0')} \leq \frac{H(\mathcal{R}_0'/\mathcal{R}_0'')}{H(\mathcal{R}_0' V \mathcal{R}_0'')} + \frac{H(\mathcal{R}_0''/\mathcal{R}_0)}{H(\mathcal{R}_0 V \mathcal{R}_0'')} . \text{ ——— } (4)$$

Adding (3) and (4) we obtain

$$\frac{H(\mathcal{R}_0/\mathcal{R}_0') + H(\mathcal{R}_0'/\mathcal{R}_0)}{H(\mathcal{R}_0 V \mathcal{R}_0')} \leq \frac{H(\mathcal{R}_0/\mathcal{R}_0'') + H(\mathcal{R}_0''/\mathcal{R}_0)}{H(\mathcal{R}_0 V \mathcal{R}_0'')} + \frac{H(\mathcal{R}_0'/\mathcal{R}_0'') + H(\mathcal{R}_0''/\mathcal{R}_0')}{H(\mathcal{R}_0' V \mathcal{R}_0'')}$$

$$\Rightarrow \quad d(\mathcal{R}_0, \mathcal{R}_0') \leq d(\mathcal{R}_0, \mathcal{R}_0'') + d(\mathcal{R}_0'', \mathcal{R}_0') . \text{ ——— } (5)$$

<u>Note 2.1.1</u>: Lemma 2.1.1 shows that the set of sub-σ-algebras, each corresponding to a discrete probability distribution, is a subset of all possible sub-σ-algebras of $\mathcal{F}$.

Thus if the probability space under consideration is $(\Omega, \mathcal{F}, \mu)$ as defined in Lemma 2.1.1, then the proof of C. Rajski's Theorem ([9]) on a metric space of discrete probability distributions follows immediately.

<u>Note 2.1.2</u>: Replacing $\mathcal{R}_0''$ by $\mathcal{R}_0 V \mathcal{R}_0'$ we obtain

$$d(\mathcal{R}_0, \mathcal{R}_0'') = \frac{H(\mathcal{R}_0/\mathcal{R}_0'') + H(\mathcal{R}_0''/\mathcal{R}_0)}{H(\mathcal{R}_0 V \mathcal{R}_0'')} = \frac{H(\mathcal{R}_0/\mathcal{R}_0 V \mathcal{R}_0') + H(\mathcal{R}_0 V \mathcal{R}_0'/\mathcal{R}_0)}{H(\mathcal{R}_0 V \mathcal{R}_0')}$$

$$= \frac{H(\mathcal{R}_0'/\mathcal{R}_0)}{H(\mathcal{R}_0 V \mathcal{R}_0')} . \text{ ——— } (6)$$

Similarly $d(\mathcal{R}_0'', \mathcal{R}_0') = \dfrac{H(\mathcal{R}_0/\mathcal{R}_0')}{H(\mathcal{R}_0 V \mathcal{R}_0')}$. ——— (7)

Adding (6) and (7), we obtain

$$d(\mathcal{R}_0, \mathcal{R}_0') = \frac{H(\mathcal{R}_0/\mathcal{R}_0') + H(\mathcal{R}_0'/\mathcal{R}_0)}{H(\mathcal{R}_0 V \mathcal{R}_0')} = d(\mathcal{R}_0, \mathcal{R}_0'') + d(\mathcal{R}_0'', \mathcal{R}_0').$$

Thus the inequality in (5) becomes equality if $\mathcal{R}_0''$ is replaced by $\mathcal{R}_0 \vee \mathcal{R}_0'$.

Corollary 2.1.1:

$$\text{If } d = \frac{H(\mathcal{R}_0/\mathcal{R}_0') + H(\mathcal{R}_0'/\mathcal{R}_0)}{H(\mathcal{R}_0 \vee \mathcal{R}_0')} \Big/ \left(1 + \frac{H(\mathcal{R}_0/\mathcal{R}_0') + H(\mathcal{R}_0'/\mathcal{R}_0)}{H(\mathcal{R}_0 \vee \mathcal{R}_0')}\right)$$

$$= \frac{H(\mathcal{R}_0/\mathcal{R}_0') + H(\mathcal{R}_0'/\mathcal{R}_0)}{H(\mathcal{R}_0 \vee \mathcal{R}_0') + H(\mathcal{R}_0/\mathcal{R}_0') + H(\mathcal{R}_0'/\mathcal{R}_0)}$$

then (Z,d) is also a bounded metric space.

Definition 2.1.1: A metric space (X,d) is convex if for any two distinct elements $x, y \in X$, there exists an element z different from both x and y and such that

$$d(x,y) = d(x,z) + d(y,z)$$

Theorem 2.1.2: If Z_1 is the sub-space of sub-σ-algebras of $\mathcal{R}$ which are such that for any two sub-σ-algebras one is not contained in the other, then (Z_1,d) is a convex metric space.

Proof: The proof immediately follows from Theorem 2.1.1, Note 2.1.2 and the Definition 2.1.1

2.2 Completeness of the Metric Space of Theorem 2.1.1

Theorem 2.2.1: The metric space (Z,d) of Theorem 2.1.1 is a complete metric space.

Proof: We are to show that any fundamental sequence $\mathcal{R}_1, \mathcal{R}_2, \ldots$ converges in Z. It is sufficient to consider the case $d(\mathcal{R}_n, \mathcal{R}_{n+p}) < \frac{1}{2^n}$ $(p > 0)$; for from any fundamental sequence we can select a subsequence satisfying this condition and a fundamental sequence that contains a convergent subsequence is convergent. We put

$$\bar{\mathcal{R}} = \bigwedge_{\ell=1}^{\infty} \bigvee_{k=\ell}^{\infty} \mathcal{R}_k$$

and show that $\bar{\mathcal{R}} \in Z$ and $d(\bar{\mathcal{R}}, \mathcal{R}_n) \longrightarrow 0$.

We prove this theorem under the assumption that $H(\mathcal{R}_0) \leq k$ (a fixed non-zero and positive constant) $\forall\ \mathcal{R}_0 \in Z$.

We have

$$d(\mathcal{R}_n, \bar{\mathcal{R}}) \leq d(\mathcal{R}_n, \bigvee_{k=n}^{\infty} \mathcal{R}_k) + d(\bigvee_{k=n}^{\infty} \mathcal{R}_k, \bar{\mathcal{R}}). \text{ —— (1)}$$

$$d(\mathcal{R}_n, \bigvee_{k=n}^{\infty} \mathcal{R}_k) = \frac{H(\mathcal{R}_n / \bigvee_{k=n}^{\infty} \mathcal{R}_k) + H(\bigvee_{k=n}^{\infty} \mathcal{R}_k / \mathcal{R}_n)}{H(\bigvee_{k=n}^{\infty} \mathcal{R}_k)}$$

$$= \frac{H(\bigvee_{k=n}^{\infty} \mathcal{R}_k / \mathcal{R}_n)}{H(\bigvee_{k=n}^{\infty} \mathcal{R}_k)} = \frac{H(\bigvee_{k=n+1}^{\infty} \mathcal{R}_k / \mathcal{R}_n)}{H(\bigvee_{k=n}^{\infty} \mathcal{R}_k)}. \text{ —— (2)}$$

Now,

$$d(\bigvee_{k=n}^{\infty} \mathcal{R}_k, \bar{\mathcal{R}}) = \frac{H(\bar{\mathcal{R}} / \bigvee_{k=n}^{\infty} \mathcal{R}_k) + (\bigvee_{k=n}^{\infty} \mathcal{R}_k / \bar{\mathcal{R}})}{H(\bigvee_{k=n}^{\infty} \mathcal{R}_k \vee \bar{\mathcal{R}})}$$

s $\bar{\mathcal{R}} \subseteq \bigvee_{k=n}^{\infty} \mathcal{R}_k$,

herefore

$$d(\bigvee_{k=n}^{\infty} \mathcal{R}_k, \bar{\mathcal{R}}) = \frac{H(\bigvee_{k=n}^{\infty} \mathcal{R}_k / \bar{\mathcal{R}})}{H(\bigvee_{k=n}^{\infty} \mathcal{R}_k)} = \frac{H(\bigvee_{k=n}^{\infty} \mathcal{R}_k) - H(\bar{\mathcal{R}})}{H(\bigvee_{k=n}^{\infty} \mathcal{R}_k)}$$

$$= 1 - \frac{H(\bar{\mathcal{R}})}{H(\bigvee_{k=n}^{\infty} \mathcal{R}_k)}. \text{ —— (3)}$$

From (1), (2) and (3) we have

$$d(\mathcal{R}_n, \bar{\mathcal{R}}) \leq \frac{H(\bigvee_{k=n+1}^{\infty} \mathcal{R}_k / \mathcal{R}_n)}{H(\bigvee_{k=n}^{\infty} \mathcal{R}_k)} + 1 - \frac{H(\bar{\mathcal{R}})}{H(\bigvee_{k=n}^{\infty} \mathcal{R}_k)}. \text{ —— (4)}$$

Now for $\ell > n$

$$H\Big(\bigvee_{k=\ell}^{\infty}\mathcal{R}_k / \bigvee_{k=n}^{\ell-1}\mathcal{R}_k\Big) = H\Big(\mathcal{R}_\ell / \bigvee_{k=n}^{\ell-1}\mathcal{R}_k\Big) + H\Big(\bigvee_{k=\ell+1}^{\infty}\mathcal{R}_k / \bigvee_{k=n}^{\ell}\mathcal{R}_k\Big)$$

$$\Rightarrow \sum_{\ell=n+1}^{\infty} H\Big(\bigvee_{k=\ell}^{\infty}\mathcal{R}_k / \bigvee_{k=n}^{\ell-1}\mathcal{R}_k\Big) = \sum_{\ell=n+1}^{\infty} H\Big(\mathcal{R}_\ell / \bigvee_{k=n}^{\ell-1}\mathcal{R}_k\Big) + \sum_{\ell=n+1}^{\infty} H\Big(\bigvee_{k=\ell+1}^{\infty}\mathcal{R}_k / \bigvee_{k=n}^{\ell}\mathcal{R}_k\Big)$$

$$\Rightarrow H\Big(\bigvee_{k=n+1}^{\infty}\mathcal{R}_k / \mathcal{R}_n\Big) = \sum_{\ell=n+1}^{\infty} H\Big(\mathcal{R}_\ell / \bigvee_{k=n}^{\ell-1}\mathcal{R}_k\Big) \leq \sum_{\ell=n+1}^{\infty} H(\mathcal{R}_\ell / \mathcal{R}_{\ell-1})$$

$$\Rightarrow \frac{H\Big(\bigvee_{k=n+1}^{\infty}\mathcal{R}_k / \mathcal{R}_n\Big)}{H\Big(\bigvee_{k=n}^{\infty}\mathcal{R}_k\Big)} \leq \frac{\sum_{\ell=n+1}^{\infty} H(\mathcal{R}_\ell / \mathcal{R}_{\ell-1})}{H\Big(\bigvee_{k=n}^{\infty}\mathcal{R}_k\Big)}$$

$$\Rightarrow \frac{H\Big(\bigvee_{k=n+1}^{\infty}\mathcal{R}_k / \mathcal{R}_n\Big)}{H\Big(\bigvee_{k=n}^{\infty}\mathcal{R}_k\Big)} \leq \frac{H(\mathcal{R}_{n+1} / \mathcal{R}_n)}{H\Big(\bigvee_{k=n}^{\infty}\mathcal{R}_k\Big)} + \frac{H(\mathcal{R}_{n+2} / \mathcal{R}_{n+1})}{H\Big(\bigvee_{k=n}^{\infty}\mathcal{R}_k\Big)} + \ldots$$

$$\Rightarrow \frac{H\Big(\bigvee_{k=n+1}^{\infty}\mathcal{R}_k / \mathcal{R}_n\Big)}{H\Big(\bigvee_{k=n}^{\infty}\mathcal{R}_k\Big)} \leq \frac{H(\mathcal{R}_{n+1} / \mathcal{R}_n)}{H(\mathcal{R}_n \vee \mathcal{R}_{n+1})} + \frac{H(\mathcal{R}_{n+2} / \mathcal{R}_{n+1})}{H(\mathcal{R}_{n+1} \vee \mathcal{R}_{n+2})} + \ldots$$

$$\Rightarrow \frac{H\Big(\bigvee_{k=n+1}^{\infty}\mathcal{R}_k / \mathcal{R}_n\Big)}{H\Big(\bigvee_{k=n}^{\infty}\mathcal{R}_k\Big)} \leq d(\mathcal{R}_n, \mathcal{R}_{n+1}) + d(\mathcal{R}_{n+1}, \mathcal{R}_{n+2}) + \ldots$$

$$< \frac{1}{2^n} + \frac{1}{2^{n+1}} + \frac{1}{2^{n+2}} + \ldots$$

$$= \frac{\frac{1}{2^n}}{1-\frac{1}{2}} = \frac{1}{2^{n-1}} \, . \qquad (5)$$

Now from (4) and (5) we obtain

$$d(\mathcal{R}_n,\bar{\mathcal{R}}) \leq \frac{1}{2^{n-1}} + 1 - \frac{H(\bar{\mathcal{R}})}{H(\bigvee_{k=n}^{\infty} \mathcal{R}_k)} .$$

Hence

$$\lim_{n\to\infty} d(\mathcal{R}_n,\bar{\mathcal{R}}) \leq \lim_{n\to\infty} \frac{1}{2^{n-1}} + 1 - \lim_{n\to\infty} \frac{H(\bar{\mathcal{R}})}{H(\bigvee_{k=n}^{\infty} \mathcal{R}_k)} .$$

$$\Rightarrow \lim_{n\to\infty} d(\mathcal{R}_n,\bar{\mathcal{R}}) \leq 0 + 1 - \frac{H(\hat{\mathcal{R}})}{H(\bar{\mathcal{R}})} \quad \text{(see [10], 5.8 - P. 16)}$$

$$\Rightarrow \lim_{n\to\infty} d(\mathcal{R}_n,\bar{\mathcal{R}}) \leq 0 \Rightarrow \lim_{n\to\infty} d(\mathcal{R}_n,\bar{\mathcal{R}}) = 0.$$

Now we prove that $H(\bar{\mathcal{R}})$ is finite.

Since $\lim_{n\to\infty} d(\bar{\mathcal{R}},\mathcal{R}_n) = 0$

we have

$$d(\bar{\mathcal{R}},\mathcal{R}_n) < \tfrac{1}{2} \text{ (say) for some } n$$

$$\Rightarrow \frac{H(\bar{\mathcal{R}}/\mathcal{R}_n) + H(\mathcal{R}_n/\bar{\mathcal{R}})}{H(\bar{\mathcal{R}} \vee \mathcal{R}_n)} < \frac{1}{2}$$

$$\Rightarrow \frac{H(\bar{\mathcal{R}}/\mathcal{R}_n)}{H(\bar{\mathcal{R}} \vee \mathcal{R}_n)} < \frac{1}{2} \Rightarrow \frac{H(\bar{\mathcal{R}} \vee \mathcal{R}_n) - H(\mathcal{R}_n)}{H(\bar{\mathcal{R}} \vee \mathcal{R}_n)} < \frac{1}{2}$$

$$\Rightarrow 1 - \frac{H(\mathcal{R}_n)}{H(\bar{\mathcal{R}} \vee \mathcal{R}_n)} < \frac{1}{2} \Rightarrow \frac{1}{2} < \frac{H(\mathcal{R}_n)}{H(\bar{\mathcal{R}} \vee \mathcal{R}_n)}$$

$$\Rightarrow H(\bar{\mathcal{R}} \vee \mathcal{R}_n) < 2H(\mathcal{R}_n) \Rightarrow H(\bar{\mathcal{R}}) \leq H(\bar{\mathcal{R}} \vee \mathcal{R}_n) < 2H(\mathcal{R}_n)$$

$$\Rightarrow H(\bar{\mathcal{R}}) \text{ is a finite constant.}$$

Note 2.2.1: Let Z be the set of equivalence classes of finite sub-σ-algebras of $\mathcal{R}$. Let $\mathcal{B}, \mathcal{C} \in Z$ and $H_{\mathcal{R}_0}(\mathcal{B} \vee \mathcal{C}) \neq 0$, then, for almost every $x \in \Omega$ metrics on the set Z may be defined as follows:

$$d^1_{\mathcal{R}_0} = d^1_{\mathcal{R}_0}(\mathcal{B},\mathcal{C}) = \frac{H_{\mathcal{R}_0}(\mathcal{B}/\mathcal{C}) + H_{\mathcal{R}_0}(\mathcal{C}/\mathcal{B})}{H_{\mathcal{R}_0}(\mathcal{B}\vee\mathcal{C})}.$$

$$d^2_{\mathcal{R}_0} = d^2_{\mathcal{R}_0}(\mathcal{B},\mathcal{C}) = \frac{H_{\mathcal{R}_0}(\mathcal{B}/\mathcal{C}) + H_{\mathcal{R}_0}(\mathcal{C}/\mathcal{B})}{H_{\mathcal{R}_0}(\mathcal{B}\vee\mathcal{C}) + H_{\mathcal{R}_0}(\mathcal{B}/\mathcal{C}) + H_{\mathcal{R}_0}(\mathcal{C}/\mathcal{B})}.$$

The othere two metrics on the set Z are:

$$d^3_{\mathcal{R}_0} = d^3_{\mathcal{R}_0}(\mathcal{B},\mathcal{C}) = \int \frac{H_{\mathcal{R}_0}(\mathcal{B}/\mathcal{C}) + H_{\mathcal{R}_0}(\mathcal{C}/\mathcal{B})}{H_{\mathcal{R}_0}(\mathcal{B}\vee\mathcal{C})}\, dP.$$

$$d^4_{\mathcal{R}_0} = d^4_{\mathcal{R}_0}(\mathcal{B},\mathcal{C}) = \int \frac{H_{\mathcal{R}_0}(\mathcal{B}/\mathcal{C}) + H_{\mathcal{R}_0}(\mathcal{C}/\mathcal{B})}{H_{\mathcal{R}_0}(\mathcal{B}\vee\mathcal{C}) + H_{\mathcal{R}_0}(\mathcal{B}/\mathcal{C}) + H_{\mathcal{R}_0}(\mathcal{C}/\mathcal{B})}\, dP.$$

2.3 Families of Metrics on the Set of Finite Sub-σ-Algebras

From Section 1.2, we know that the generalized conditional entropy $I_\alpha(\mathcal{R}_0'/\mathcal{R}_0)$ for the finite sub-σ-algebras $\mathcal{R}_0$ and $\mathcal{R}_0'$ of $\mathcal{R}$ satisfies the properties (i) to (vi). Properties (iii) to (vi) are true for $\alpha \geq 0$ and (i) and (ii) are valid for $\alpha > 0$ and $\alpha \geq 1$ respectively.

Now we define two new metrics on the set Z of all finite sub-σ-algebras of $\mathcal{R}$ as follows:

$$d^1_\alpha = d^1_\alpha(\mathcal{R}_0,\mathcal{R}_0') = I_\alpha(\mathcal{R}_0/\mathcal{R}_0') + I_\alpha(\mathcal{R}_0'/\mathcal{R}_0); \quad \mathcal{R}_0, \mathcal{R}_0' \in Z,$$

$$d^2_\alpha = d^2_\alpha(\mathcal{R}_0,\mathcal{R}_0') = \frac{I_\alpha(\mathcal{R}_0/\mathcal{R}_0') + I_\alpha(\mathcal{R}_0'/\mathcal{R}_0)}{I_\alpha(\mathcal{R}_0\vee\mathcal{R}_0')}; \quad \mathcal{R}_0, \mathcal{R}_0' \in Z$$

$$I_\alpha(\mathcal{R}_0\vee\mathcal{R}_0') \neq 0.$$

It can be easily verified that (Z, d^1_α) is a metric space for all $\alpha \geq 1$ and (Z, d^2_α) is a metric space for all $\alpha \geq 1$.

<u>Note 2.3.1</u>: For $\alpha = 1$, $I_\alpha(\mathcal{R}_0'/\mathcal{R}_0)$ reduces to Shannon's conditional entropy. Thus if Z is the set of equivalence classes of finite sub-σ-algebras of $\mathcal{R}$, then

$$D_1 = \{d_\alpha^1;\ \alpha \in [1,\infty)\} \text{ and } D_2 = \{d_\alpha^2;\ \alpha \in [1,\infty)\}$$

represent two families of metrics on Z such that the metric given $d = H(\mathcal{R}_0/\mathcal{R}_0') + H(\mathcal{R}_0'/\mathcal{R}_0)$ belongs to D_1 and the metric given by $d = \dfrac{H(\mathcal{R}_0/\mathcal{R}_0') + H(\mathcal{R}_0'/\mathcal{R}_0)}{H(\mathcal{R}_0 \vee \mathcal{R}_0')}$ belongs to D_2.

The author wishes to express his sincere appreciation to Dr. M. Behara for his willing assistance and guidance and for the generosity with which he gave his valuable time during the course of this research.

References

1. Behara, M. and Nath, P. (1971), Additive and Non-additive Entropies of Finite Measurable Partitions, Lecture notes in Mathematics, Probability and Information Theoy II, Springer-Verlag.

2. Billingsley, P., Ergodic Theory and INformation, New York, Wiley 1965.

3. Brown, T.A. (1963), Entropy and Conjugacy, Ann. Math. Stat., 34, 226-232.

4. Copson, E.T., Metric Spaces, Cambridge University Press - 1968.

5. Jacobs, K., Lecture Notes on Ergodic Theory, Part II, Matematisk Institut, Aarhus Universitet.

6. Loeve, M., Probability Theory, New York, Van Nostrand 1955.

7. Neveu, J., Mathematical Foundations of the Calculas of Probability, Holden-Day, Inc., San Francisco, London, Amsterdam.

8. Nijst, A.G.P.M., Some Remarks on Conditional Entropy, 22 Z. Wahrscheinlichkeitstheorie Verw. Geb; Bd. 12, P. 307-319.

9. Rajski, C., A Metric Space of Discrete Probability Distributions, Information and Control 4, 371-377 (1961).

10. Rokhlin, V.A., Lectures on the Entropy Theory of Measure Preserving Transformations, Russ. Math. Surveys 22, 1-52, (1967).

11. Royden, H.L., Real Analysis, Second Edition, The MacMillan Company, New York, Collier-MacMillan Limited, London.

12. Singh, J.M., Metrization of Sets of Sub-σ-Algebras and their Conditional Entropies, Thesis, Department of Mathematics, McMaster University, Canada.

Groups with Chu Duality

Herbert Heyer

Tübingen (Germany)

Contents

1. Introduction

The first attempt of presenting an axiomatic theory of duality in topological groups has been successfully made by H. Chu in 1966 in his interesting paper [1]. Chu introduces a general duality principle and investigates classes of groups which satisfy this principle. The principle is set up to include the most familiar duality theorems of T. Tannaka for compact groups and L. S. Pontrjagin for locally compact abelian groups. Proofs of both theorems can be found in various books. The reader is invited to consult either the standard encyclopedia on harmonic analysis [10] or a more modest account on the subject like the author's Lecture Notes [9].

Meanwhile the representation theory for arbitrary locally compact groups has been developed so immensely that it seems fruitful to investigate further classes of locally compact groups with respect to their property of satisfying Chu's duality principle. The classes under consideration are those which contain compact groups as well as locally compact abelian groups as special cases. Here the most attractive class seems to be the class of Moore groups. Moore groups have been characterized by C. C. Moore [13] and L. Robertson [15].
Their recent results imply for example that every Moore group satisfies the duality principle, a result which under technical restrictions had been proved in [9].
On the other hand it turned out that not every locally compact group satisfying the duality principle is a Moore group.
At the time of publication of [9] Moore's paper had not been available and the proofs in [15] were hardly understandable without knowing the methods used by Moore. This was the reason why the author did not include the most recent results in [9]. The aim of this paper is to provide a supplement to [9] on the basis of [13] and [15] including

new properties of groups satisfying the duality principle and to discuss the interplay between various classes of groups that are evidently connected with the duality theorems available. In general the proofs of the theorems have just been sketched (the arguments being given in brackets). Detailed demonstrations of the known facts can be found either in [9] or in the original papers quoted therein.

The author is indebted to Mr. E. Siebert for adding a few interesting comments to the paper.

2. Chu dual and Chu duality

Let G be an arbitrary topological group. For each $n \geq 1$ one denotes by $Rep_n(G)$ the totality of (continuous, unitary) representations D of G with n-dimensional representation space $\mathcal{H}(n) := \mathcal{H}(n(D))$, where n(D) is the dimension of D.

In $Rep_n(G)$ the compact open topology $\mathcal{T}_c$ is introduced in the following fashion: For each $D \in Rep_n(G)$, each compact subset K of G and each $\varepsilon > 0$ one defines the set

$$W(D;K,\varepsilon) := \{D' \in Rep_n(G) : \| D'(x)-D(x)\| < \varepsilon \text{ for all } x \in K\},$$

the collection

$$\mathcal{W}(D) := \{W(D;K,\varepsilon) : K \text{ compact in } G, \varepsilon > 0\}$$

and takes the family

$$\mathcal{W} := \{\mathcal{W}(D) : D \in Rep_n(G)\}$$

as a base for the topology in $Rep_n(G)$.

Let Rep(G) be the (disjoint) topological sum of the spaces $Rep_n(G)$ $(n \geq 1)$.

[In fact one can always achieve $Rep_m(G) \subset Rep_n(G)$ for $m \leq n$ by considering the mapping $j_{mn} : Rep_m(G) \to Rep_n(G)$ defined by

$$j_{mn}(D) := D \oplus I_{n-m} \text{ for all } D \in Rep_m(G).]$$

Then Rep(G) is called the <u>Chu dual</u> of G.

By $Irr_n(G)$ $(n \geq 1)$ and Irr(G) we denote the subspaces consisting of all irreducible representations in $Rep_n(G)$ and Rep(G) resp.

Properties of Rep(G)

2.1 If G is locally compact, then Rep(G) is locally compact and uniformizable.

[The proof of local compactness of Rep(G) is deduced from a more general result of M. Goto [5] : Let G be a locally compact group and let L be a compact Lie group. Then the set Hom(G,L) of all continuous homomorphisms from G into L is locally compact w.r.t. the topology $\mathcal{T}_c$ in Hom(G,L).]

2.2 If G is a discrete (topological) group, then $Rep_n(G)$ is compact for every $n \geq 1$ (and thus Rep(G) is σ-compact).

[It is enough to show that $Rep_n(G)$ is compact for each $n \geq 1$ and this is done by embedding $Rep_n(G)$ homeomorphically onto a closed subgroup of $\mathfrak{U}_o := \prod_{x \in G} \mathfrak{U}_x(n)$ with $\mathfrak{U}_x(n)$ equal to the unitary group $\mathfrak{U}(n)$ for all $x \in G$.

2.3 Let G be a second countable locally compact group. Then also Rep(G) is second countable (locally compact).

[The proof follows the idea of Pontrjagin developed for the abelian case.]

2.4 If G is a second countable locally compact group, then Rep(G) and Irr(G) are polish.

[It suffices to show that $Irr_n(G)$ is a G_δ-set in $Rep_n(G)$ $(n \geq 1)$. For each $\eta \in \mathcal{H}(n) \setminus \{0\}$ and each $r \in \mathbb{R}$ with $0 < r < \|\eta\|$ one defines the closed r-ball in $\mathcal{H}(n)$ with center η by $K(\eta, r)$ and puts

$$O(\eta,r) := \{D \in \mathrm{Rep}_n(G) : \langle D(G)\xi \rangle = \mathcal{H}(n) \text{ for all } \xi \in K(\eta,r)\}$$

Then $\mathrm{Irr}_n(G)$ can be written as a countable intersection of open sets of the form $O(\eta,r)$.

For $n \geq 1$ let $\mathrm{Rep}_n(G)/\sim$ be the set of (unitary) equivalence classes of elements of $\mathrm{Rep}_n(G)$ furnished with the quotient topology.

2.5 If G is compact, then $\mathrm{Rep}_n(G)/\sim$ is discrete for every $n \geq 1$.

[One shows that for a compact group G to every $D \in \mathrm{Rep}_n(G)$ there exists a neighborhood $W := W(D;G,\frac{1}{n})$ such that $D' \sim D$ for all $D' \in W$.]

For any topological group G and any closed subgroup H of G one defines the orthogonal complement or (first) annihilator of H w.r.t. $\mathrm{Rep}_n(G)$ ($n \geq 1$) or $\mathrm{Rep}(G)$ resp. by

$$H_n^{\perp} := [\mathrm{Rep}_n(G),H] = \{D \in \mathrm{Rep}_n(G) : D(z) = I_n \text{ for all } z \in H\}$$

or

$$H^{\perp} := [\mathrm{Rep}(G),H] = \{D \in \mathrm{Rep}(G) : D(z) = I_{n(D)} \text{ for all } z \in H\}.$$

The elementary properties of the annihilator are proved just as in the theory of locally compact abelian groups.

Let G be a locally compact group. Then

2.6 $H_n^{\perp}$ is closed in $\mathrm{Rep}_n(G)$ ($n \geq 1$).

2.7 If H is a normal subgroup of G, then $H_n^{\perp}$ is homeomorphic to $\mathrm{Rep}_n(G/H)$ ($n \geq 1$) and thus $H^{\perp}$ homeomorphic to $\mathrm{Rep}(G/H)$.

[Let p denote the canonical mapping from G onto G/H and for every $D \in H_n^{\perp}$ let $D_H \in \mathrm{Rep}_n(G/H)$ be defined by $D := D_H \circ p$. Then the mapping f defined by $f(D) := D_H$ for all $D \in H_n^{\perp}$ is the desired homeomorphism.]

For any closed normal subgroup of the topological group G we define the <u>second annihilator</u> of H by

$$H^{\perp\perp} := [G, H^{\perp}] = \{ x \in G : D(x) = I_{n(D)} \text{ for all } D \in H^{\perp} \}$$

2.8 If G is a locally compact group and G/H is maximally almost periodic, then $H^{\perp\perp} = H$.

[To be shown is $H^{\perp\perp} \subset H$. If there is an $a \in H^{\perp\perp} \setminus H$, then one has for the mapping f defined in the remark following 2.7: $f(D)(aH) = I_{n(D)}$ for all $D \in H^{\perp}$ which contradicts the maximal almost periodicity of G/H.]

2.9 Let G be a locally compact maximally almost periodic group and let $Rep_n(G)$ be connected for every $n \geq 1$. Then G has no compact subgroup $\neq \{e\}$.

[Let H be a compact subgroup of G and $x \in H$ with $x \neq e$. Then for every $n \geq 1$ $H_n^{\perp}$ is closed and by the remark following 2.5 also open in $Rep_n(G)$. Thus $H_n^{\perp} = Rep_n(G)$, i.e. $D(x) = I_{n(D)}$ for all $D \in Rep_n(G)$ which contradicts the maximal almost periodicity of G.]

We proceed to the definition of the second dual object attached to a topological group G.

<u>Definition</u>: A <u>quasi representation</u> of G is any continuous mapping Q from Rep(G) into the topological sum $\mathfrak{U} := \bigcup_{n \geq 1} \mathfrak{U}(n)$ of all unitary groups $\mathfrak{U}(n)$ $(n \geq 1)$ with the following properties:

(Q1) $Q(D) \in \mathfrak{U}(n(D))$ for all $D \in Rep(G)$

(Q2) $Q(D \oplus D') = Q(D) \oplus Q(D')$ for all $D, D' \in Rep(G)$

(Q3) $Q(D \otimes D') = Q(D) \otimes Q(D')$ for all $D, D' \in Rep(G)$

(Q4) $Q(U^{-1}DU) = U^{-1}Q(D)U$ for all $D \in \mathrm{Rep}(G)$ and
$U \in \mathfrak{U}(n(D))$

Let $\mathrm{Rep}(G)^{\vee}$ be the totality of quasi representations of G. Introducing in $\mathrm{Rep}(G)^{\vee}$ the topology $\mathcal{T}_c$ (with resepct to the strong topology in $\mathfrak{U}$) one obtains a topological space called the Chu quasi dual of G.

Properties of $\mathrm{Rep}(G)^{\vee}$

2.10 $\mathrm{Rep}(G)^{\vee} \neq \emptyset$.

2.11 For each $Q \in \mathrm{Rep}(G)^{\vee}$ one has $\overline{Q(D)} = Q(\overline{D})$ for all $D \in \mathrm{Rep}(G)$.

2.12 For the identity representation $E \in \mathrm{Rep}(G)$ one has $Q(E) = I$ $(Q \in \mathrm{Rep}(G)^{\vee})$

[From (Q4) one deduces $Q(E) = \alpha I$ for $\alpha \in \mathbb{C}^*$, but for each $n \geq 1$ one obtains $Q(E_{n^2}) = Q(nE_n)$ using (Q2) and (Q3) and thus $\alpha = 1$.]

In order to introduce a group structure in $\mathrm{Rep}(G)^{\vee}$ one defines for any two elements Q, Q' of $\mathrm{Rep}(G)^{\vee}$ their product QQ' by

$$QQ'(D) := Q(D)Q'(D)$$

for all $D \in \mathrm{Rep}(G)$.

Putting in addition for a given $Q \in \mathrm{Rep}(G)^{\vee}$

$$Q^{-1}(D) := (Q(D))^{-1}$$

and

$$E(D) := I_{n(D)}$$

for all $D \in \mathrm{Rep}(G)$, $\mathrm{Rep}(G)^{\vee}$ becomes a group with E as its identity.

2.13 $Rep(G)^{\vee}$ is a Hausdorff (topological) group.

[The Hausdorff property is obviously satisfied. One verifies without efforts that the mappings $(Q,Q') \to QQ'$ and $Q \to Q^{-1}$ from $Rep(G)^{\vee} \times Rep(G)^{\vee}$ and $Rep(G)^{\vee}$ into $Rep(G)^{\vee}$ resp. are continuous.]

Remark: It is not known whether in general $Rep(G)^{\vee}$ is locally compact.

2.14 $Rep(G)^{\vee}$ is maximally almost periodic.

[Let $D \in Rep(G)$ and put $\hat{D}(Q) := Q(D)$ for all $Q \in Rep(G)^{\vee}$. Then $\hat{D}$ is a (finite dimensional) representation of $Rep(G)^{\vee}$ which can be used to separate the points of $Rep(G)^{\vee}$.]

The following theorem contains the general information on the connection between $Rep(G)^{\vee}$ and the underlying group G.

For every $x \in G$ one defines $\check{x}$ by

$$\check{x}(D) := D(x)$$

for all $D \in Rep(G)$.

Plainly $\check{x} \in Rep(G)^{\vee}$.

Let the mapping Ω from G into $Rep(G)^{\vee}$ be defined by

$$\Omega(x) := \check{x}$$

for all $x \in G$.

Theorem 2.1 For any topological group G with Chu quasi dual $Rep(G)^{\vee}$ the above defined mapping Ω is a homomorphism, which is continuous, if G is locally compact, and injective iff G is maximally almost periodic.

[It remains to show that Ω is continuous, but this follows from the continuity of the mapping g from $G \times \mathrm{Rep}_n(G)$ into $\mathfrak{U}(n)$ defined by $g(x,D) := \Omega(x)(D) = D(x)$ for all $x \in G$ and $D \in \mathrm{Rep}_n(G)$ $(n \geq 1)$.]

Remark: In general Ω defined in theorem 2.1 is not a topological isomorphism. This follows from 2.13 since there exist non maximally almost periodic groups. See also the example in §4 below.

The remark motivates the following

Definition: Let G be a topological group and $\mathrm{Rep}(G)^{\vee}$ its Chu quasi dual. Then G is said to be a group with Chu duality (or one says that G has Chu duality), if Ω defined in theorem 2.1 is in fact a topological isomorphism.

In general Ω is called the Chu homomorphism of G. If G has Chu duality, then Ω is called the Chu duality of G.

Let us denote by $\mathfrak{C}$ the class of all (locally compact) groups with Chu duality.

The aim of the paper is to study this class $\mathfrak{C}$, exhibit properties of $\mathfrak{C}$ and provide most general examples of groups in $\mathfrak{C}$.

Examples: (1) Every compact group is in $\mathfrak{C}$ (and the Chu duality does the same as the well-known Tannaka duality).
(2) Every locally compact abelian group is in $\mathfrak{C}$ (and the Chu duality does the same as the Pontrjagin duality).
(3) Every Takahashi group G, that is a locally compact maximally almost periodic group G with compact commutator subgroup K(G), is in $\mathfrak{C}$. In particular every central group, that is a locally compact group G such that the quotient G/Z(G) by the center Z(G) of G is compact, is in $\mathfrak{C}$.

[The proof is achieved by invoking the Takahashi duality theorem [17]: See [9], 346. The particular case follows on the basis of [6], 331]

More general examples will be discussed in §4. We finally illustrate what the Chu quasi dual looks like in the case of a Bohr compactification. To each topological group G there corresponds its Bohr compactification $(\tilde{G},\varphi)$, where $\tilde{G}$ is a compact group and φ is a continuous homomorphism from G on a dense subgroup of $\tilde{G}$ with the universal mapping property.

2.15 For any topological group G let $\mathrm{Rep}(G)_d$ be the discretely topologized (Chu) dual of G. Then $(\mathrm{Rep}(G)_d)^{\vee}$ is topologically isomorphic to $\tilde{G}$.

[Plainly $\mathrm{Rep}(G)_d$ and $\mathrm{Rep}(\tilde{G})_d$ are topologically isomorphic. The Chu duality for the compact group $\tilde{G}$ (which coincides with the Tannaka duality) yields the desired topological isomorphism.]

2.16 For every locally compact group G the group $(\mathrm{Rep}(G)^{\vee})^{\sim}$ is topologically isomorphic to $\tilde{G}$.

[This follows immediately from the uniqueness of the Bohr compactification.]

3. Properties of groups with Chu duality

We start with two minor properties and proceed to the closure properties of the class $\mathcal{C}$ later.

3.1 Let $G \in \mathcal{C}$. Then $\mathrm{Rep}(G)/\sim$ is discrete iff G is compact.

[By 2.5 it suffices to show that the discreteness of $\mathrm{Rep}(G)/\sim$ implies the compactness of G.

We note that the topology of $Rep(G)^{\vee}$ in this case is equal to the topology of pointwise convergence on $Rep(G)/\sim$. The mapping ϕ from $Rep(G)^{\vee}$ into $\mathfrak{U}_1 := \prod_{D \in Rep(G)} \mathfrak{U}(n(D))$ defined by

$$\phi(Q) := (D(x))_{D \in Rep(G)}$$

for all $Q = \check{x} \in Rep(G)^{\vee}$ is a homeomorphism onto $\phi(Rep(G)^{\vee})$ which is a compact subgroup of $\mathfrak{U}_1$.

For any topological group G and any $n \geq 1$ we define the von Neumann kernel $G^{(n)}$ for $Rep_n(G)$ by

$$G^{(n)} := \bigcap_{D \in Rep_n(G)} \ker D$$

Obviously $G^{(n)}$ is a closed normal subgroup of G for every $n \geq 1$ and, of course:

$$G^{\circ} := \bigcap_{n \geq 1} G^{(n)}$$

is the von Neumann kernel for Rep(G) or of G as it was introduced in [9], 175.

3.2 Let $G \in \mathcal{C}$ and let V be a neighborhood of the identity e in G. Then there exists an $n \geq 1$ such that

$$G^{(n)} \subset V$$

[Let Ω be the Chu duality of G. Then $\Omega(V)$ is a neighborhood of the identity in $Rep(G)^{\vee}$. For any $m \geq 1$, compact subsets K of $Rep_m(G)$ and open neighborhoods W of the identity in $\mathfrak{U}(m)$ of the form $W := \{ T \in \mathfrak{U}(m) : \| T - I_m \| < \varepsilon \}$ we define

$$N(K,W) := \{ Q \in Rep(G)^{\vee} : Q(D) \in W \text{ for all } D \in K \}.$$

$Rep(G)^\vee$ carries the topology $\mathcal{T}_c$. Thus there is an $m \geq 1$ and for any $i = 1,\dots,k$ there are a compact subset K_i of $Rep_m(G)$ and an open neighborhood W_i of the identity in $\mathfrak{U}(m)$ such that

$$\bigcap_{i=1}^{k} N(K_i, W_i) \subset \Omega(V)$$

Choosing $n \geq 1$ sufficiently large such that for $i=1,\dots,k$

$$K_i \subset Rep_m(G)$$

holds with $m \leq n$ one obtains for any $x \in G^{(n)}$:

$$\Omega(x) \in \Omega(V)$$

and, since Ω is injective, the result.]

In the following we will be concerned with topological groups G, G_1 and G_2 and sequences

(a) $\quad \{e\} \longrightarrow G_1 \xrightarrow{j} G \xrightarrow{p} G_2 \longrightarrow \{e\}$

(b) $\quad \{E\} \longleftarrow Rep(G_1) \xleftarrow{j^\wedge} Rep(G) \xleftarrow{p^\wedge} Rep(G_2) \longleftarrow \{E\}$

(c) $\quad \{E\} \longrightarrow Rep(G_1)^\vee \xrightarrow{j^\vee} Rep(G)^\vee \xrightarrow{p^\vee} Rep(G_2)^\vee \longrightarrow \{E\}$

where E denotes the identity representation as well as the identity quasi representation of the underlying groups G_1 and G_2 and where the mappings j,p are just continuous homomorphisms, $j^\wedge$, $p^\wedge$, $j^\vee$ and $p^\vee$ are defined via j and p in the following manner:

$$j^\wedge(D) := D \circ j \quad \text{for all } D \in Rep(G)$$

$$p^\wedge(D) := D \circ p \quad \text{for all } D \in Rep(G_2)$$

$$j^\vee(Q) := Q \circ j^\wedge \quad \text{for all } Q \in Rep(G_1)^\vee \quad \text{and}$$

$$p^\vee(Q) := Q \circ p^\wedge \quad \text{for all } Q \in Rep(G)^\vee$$

We will consider a few special cases of groups G, G_1 and G_2.

3.3 Let $(G_\iota)_{\iota \in I}$ be a family of locally compact groups G_ι with a filtered index set I and $\varprojlim_{\iota \in I} G_\iota$ its projective limit. Replacing in the above sequences G by $\varprojlim_{\iota \in I} G_\iota$, G_1 by ker p_ι where p_ι is the projection of G onto G_ι and G_2 by G_ι $(\iota \in I)$ we can proceed as follows: Let for each $\iota \in$ I the group G_ι have Chu duality, that is for $\iota \in$ I and $Q \in \mathrm{Rep}(G_\iota)^\vee$ there is an $x_\iota \in G_\iota$ such that

$$\check{x}_\iota = \Omega(x_\iota) = Q$$

Let now $Q \in \mathrm{Rep}(G)^\vee$. Then

$$\check{p}_\iota(Q) \in \mathrm{Rep}(G_\iota)^\vee$$

thus there exists an $x = (x_\iota)_{\iota \in I} \in \prod_{\iota \in I} G_\iota$ which turns out to be in G such that

$$\check{p}_\iota(Q) = \check{x}_\iota$$

holds. For $D \in \mathrm{Rep}(G)$ there exists a $D_\iota \in \mathrm{Rep}(G_\iota)$ with $D = D_\iota \circ p_\iota$ for all $\iota \in$ I and therefore one obtains

$$Q(D) = Q(D_\iota \circ p_\iota) = \check{p}_\iota(Q)(D) = D_\iota(x_\iota) = \check{x}(D_\iota \circ p_\iota) = \check{x}(D) = \Omega(x)$$

Thus Ω is surjective.

Further more Ω^{-1} is continuous, thus G has Chu duality.

We proved the following

<u>Theorem 3.1</u> $\mathcal{C}$ is closed under forming projective limits.

3.4 Let G_1 and G_2 be locally compact groups and $G: = G_1 \times G_2$ their direct product. Then with injection j and projection p one obtains a topological isomorphism between $\mathrm{Rep}(G)^{\vee}$ and $\mathrm{Rep}(G_1)^{\vee} \times \mathrm{Rep}(G_2)^{\vee}$ and an exact sequence (c).

[For a proof see [1], 317 or [9], 355.]

Applying the Five Lemma of algebraic topology to this fact one arrives at the following

<u>Theorem 3.2</u> $\mathcal{L}$ is closed under forming (finite) direct products.

3.5 Let G be a topological group and H a closed normal subgroup of G with $[G:H] < \infty$. Then one can introduce for every $D \in \mathrm{Rep}(H)$ its induced representation Ind(D) in Rep(G).

In fact: Given $D \in \mathrm{Rep}(H)$ (with representation space $\mathcal{H}$) of the form $D = (d_{ij})_{i,j=1,\dots,m}$ (w.r.t. a fixed base in $\mathcal{H}$) and a group G of the form $\bigcup_{j=1}^{n} a_j H$ with $a_1 = e$ one defines

$$F(x) := \begin{cases} (d_{ij}(x))_{i,j=1,\dots,m} & \text{for } x \in H \\ 0 & \text{otherwise} \end{cases}$$

and for all $x \in G$:

$$\Delta_{hk}(x) := F(a_h^{-1} x a_k)$$

whenever $h,k = 1,\dots,n$.

Then $\mathrm{Ind}(D) := (\Delta_{hk})_{h,k=1,\dots,n}$.

The following properties are easily established:

(1) Let $D_1, D_2 \in \mathrm{Rep}(H)$ with $D_1 \sim D_2$. Then $\mathrm{Ind}(D_1) \sim \mathrm{Ind}(D_2)$.

(2) Let $D \in \mathrm{Rep}(H)$. Then $\mathrm{Res}_H(\mathrm{Ind}(D)) = D^{(1)} \oplus \dots \oplus D^{(m)}$, where for every

$h = 1,\dots,m$ $D^{(h)}$ is defined by

$$D^{(h)}(x) := D(a_h^{-1} x a_h)$$

for all $x \in H$.

(3) Let $D_1, D_2 \in \mathrm{Rep}(H)$. Then
$\mathrm{Ind}(D_1 \oplus D_2) = \mathrm{Ind}(D_1) \oplus \mathrm{Ind}(D_2)$.

(4) Let $D_1 \in \mathrm{Rep}(H)$ and $D \in \mathrm{Rep}(G)$. Then
$\mathrm{Ind}(D_1 \otimes \mathrm{Res}_H D) \sim \mathrm{Ind}(D_1) \otimes D$.

(5) Let $D_1, D_2 \in \mathrm{Rep}(H)$. Then
$\mathrm{Ind}(D_1 \otimes D_2^{(1)}) \oplus \dots \oplus \mathrm{Ind}(D_1 \otimes D_2^{(m)}) \sim \mathrm{Ind}(D_1) \otimes \mathrm{Ind}(D_2)$

Let $Q \in \mathrm{Rep}(G)^{\vee}$ and $D \in \mathrm{Rep}(H$.

We define $Q_o(D)$ as the $n(D) \times n(D)$-submatrix in the left upper corner of $Q(\mathrm{Ind}(D))$.

Let $D_1, D_2 \in \mathrm{Rep}(H)$. Then one has

(6) $Q_o(D_1 \oplus D_2) = Q_o(D_1) \oplus Q_o(D_2)$ and

(7) $Q_o(D_1 \otimes D_2) = Q_o(D_1) \otimes Q_o(D_2)$.

If $D \in \mathrm{Rep}(H)$ and $U \in \mathfrak{U}(n(D))$, then

(8) $Q_o(U^{-1}DU) = U^{-1}Q_o(D)U$.

We now apply the notation of the sequences (a), (b) and (c) to $G_1 := H$ and $G_2 := G/H$ (G itself remaining the same) and to the injection j from H into G and the projection p from G onto G/H.

(9) Let $Q \in \ker p^{\vee}$. Then for every $D \in \mathrm{Rep}(H)$
$Q_o(D)$ is a direct summand of $Q(\mathrm{Ind}(D))$.

[The proofs of the properties (1) to (9) can be translated from [11] into our situation.

We present the reasoning for (9):

In fact by (4) we have for every $D \in Rep(H)$

$$Ind(E) \otimes Ind(D) \sim Ind(Res_H(Ind(D)))$$
$$= Ind(D^{(1)} \oplus \ldots \oplus D^{(m)})$$

Let now $Q \in \ker p^{\vee}$. Then we obtain

$$\underbrace{Q(D) \oplus \ldots \oplus Q(D)}_{m\text{-times}} = E \otimes Q(Ind(D))$$
$$\sim Q(Ind(D^{(1)} \oplus \ldots \oplus D^{(m)}))$$

Therefore $Q(Ind(D)) = Q_o(D^{(1)} \oplus \ldots \oplus D^{(m)})$

$$= Q_o(D^{(1)}) \oplus \ldots \oplus Q_o(D^{(m)})$$

(using (6)) and $Q_o(D) = Q_o(D^{(1)})$ is a direct summand of $Q(Ind(D))$.

We proceed to the study of the duality properties of the following diagram with a locally compact group G and a closed normal subgroup H of G with $[G:H]<\infty$:

$$\begin{array}{llcccccccc}
(a) & \{e\} & \longrightarrow & H & \xrightarrow{j} & G & \xrightarrow{p} & G/H & \longrightarrow & \{e\} \\
 & & & \Omega_1 \downarrow & & \Omega \downarrow & & \Omega_2 \downarrow & & \\
(c) & \{E\} & \longrightarrow & Rep(H)^{\vee} & \xrightarrow{j^{\vee}} & Rep(G)^{\vee} & \overset{p^{\vee}}{\text{——}} & Rep(G/H)^{\vee} & \longrightarrow & \{E\}
\end{array}$$

where Ω_1, Ω and Ω_2 are the Chu homomorphisms of H,G and G/H resp.

Let now $H \in \mathcal{C}$. Of course $G/H \in \mathcal{C}$, since it is finite. Therefore Ω_1 and Ω_2 are topological isomorphisms. Obviously $j^{\vee}$ and $p^{\vee}$ are continuous homomorphisms.

1. The diagram is commutative.

It suffices to show the commutativity for the two inner squares, but this follows easily:

For all $x \in H$ and $D \in Rep(G)$ one has

$$\Omega \circ j(x)(D) = D(j(x)) = \Omega_1(x)(Res_H(D)) = (j^{\vee} \circ \Omega_1)(x)(D)$$

and for all $x \in G$ and $D \in Rep(G/H)$:

$$\Omega_2 \circ p(x)(D) = D(xH) = \Omega(x)(D'), \text{ where } D'(x) = D(xH)$$

for all $x \in G$, and thus

$$\Omega_2 \circ p(x)(D) = (p^{\vee} \circ \Omega)(x)(D).$$

2. The sequences (a) and (c) are exact.

We just have to prove the exactness of (c):

With the notions of the above proof we have for every $D \in Rep(G/H)$: $Res_H D' = E$, thus

$$im\ j^{\vee} \subset ker\ p^{\vee}$$

Let now $Q \in ker\ p^{\vee}$. Then for each $D \in Rep(H)$ $Q_o(D)$ is a direct summand of $Q(Ind(D))$ by (9). But this implies $Q \in im\ j^{\vee}$, since $j^{\vee}(Q_o) = Q$ for $Q_o \in Rep(H)^{\vee}$. Thus $im\ j^{\vee} = ker\ p^{\vee}$.

$j^{\vee}$ is injective: In fact let $Q \in ker\ j^{\vee}$, i.e. $Q(Res_H D) = I_{n(Res_H D)}$ for all $D \in Rep(G)$. Then for all $D_1 \in Rep(H)$ one has $I_{n(D_1)}$ $= Q(Res_H(Ind(D_1))) = Q(D_1^{(1)} \oplus \ldots \oplus D_1^{(m)})$ by (2), thus $Q(D_1) = Q(D_1^{(1)})$ $= I_{n(D_1)}$.

$p^{\vee}$ is surjective: In fact: G/H being a group with Chu duality we obtain:

$$Rep(G/H)^{\vee} = im\ \Omega_2 = im(\Omega_2 \circ p) = im(p^{\vee} \circ \Omega) \subset im\ p^{\vee} \subset Rep(G/H)^{\vee}\ ]$$

We have established that Ω is a continuous injective homomorphism which by the Five Lemma of algebraic topology is a continuous isomorphism.

3. Ω is open.

[It will be sufficient to show that $j^{\vee}$ is open, since Ω will be open iff $\mathrm{Res}_H \Omega$ is open.

Let $\varepsilon \in]0,1[$ and K be a compact subset of Rep(H). Then

$$U(K,\varepsilon) := \{\Omega(y) : y \in H \text{ and } \|D(y)-I_{n(D)}\| < \varepsilon \text{ for all } D \in K\}$$

is a neighborhood of the identity E in $\mathrm{Rep}(H)^{\vee}$.

Obviously the mapping $D \longrightarrow \mathrm{Ind}(D)$ from Rep(H) into Rep(G) is continuous, thus

$$K^* := \{\mathrm{Ind}(D) : D \in K\}$$

is compact in Rep(G).

We show that

$$U(K^*,\varepsilon) := \{\Omega(x) : \|\mathrm{Ind}(D)(x)-I_{n(\mathrm{Ind}(D))}\| < \varepsilon \text{ for all } D \in K\}$$

is contained in $j^{\vee}(U(K,\varepsilon))$.

Plainly for $y \in H$ and $\Omega(y) \in U(K^*,\varepsilon)$ one has

$$(*) \quad \|D(y)-I_{n(D)}\| \leq \|\mathrm{Ind}(D)(y)-I_{n(\mathrm{Ind}(D))}\| < \varepsilon$$

for all $D \in K$.

Furthermore for $x \in G \setminus H$

$$\|Ind(D)(x) - I_{n(Ind(D))}\| \geq \sqrt{n(Ind(D))} \geq 1 > \varepsilon$$

holds whenever $D \in K$.

Thus $\Omega(x) \in U(K^*, \varepsilon)$ implies $x \in H$ and by (*)

$$\Omega(x) \in j^\vee \quad (U(K, \varepsilon)).$$

Since $j^\vee$ is a homomorphism, it is clearly open.

We obtained

Theorem 3.3 $\mathcal{L}$ is closed under forming finite extensions.

Corollary 3.1 For any closed subgroup H of finite index in a locally compact group G one has $H \in \mathcal{L}$ iff $G \in \mathcal{L}$.

To be proved: If G and H is a closed normal subgroup of finite index in G or more generally if $G \in \mathcal{L}$ and H is a closed normal subgroup of G such that G/H is maximally almost periodic, then $H \in \mathcal{L}$.
Let $Q \in Rep(H)^\vee$ be given. To $Q' := j(Q)$ there exists an $x \in G$ such that

$$Q'(D) = \check{x}(D) = D(x)$$

for all $D \in Rep(G)$.
Assume that $x \in H$. Then there is a $D_1 \in Rep(G/H)$ such that $D_1(xH) \neq D_1(H) = I_{n(D_1)}$ and thus a $D \in Rep(G)$ such that $D \neq E$ but $D(y) = I_{n(D)}$ for all $y \in H$.
Then we obtain for the above $x \in H$:

$$I_{n(D)} \neq D(x) = Q'(D) = j^\vee(Q)(D) = Q(D \circ j) = I_{n(D)}$$

which yields a contradiction. Therefore $x \in H$.

Thus the Chu homomorphism Ω_H of H is surjective. Since the Chu homomorphism Ω of G is open and since $j^\vee \circ \Omega_H = \Omega \circ j$ holds, Ω_H itself is open.

<u>Corollary 3.2</u> For any closed subgroup H of finite index in a locally compact group G one has $H \in \mathcal{L}$ iff $G \in \mathcal{L}$

[Follows from corollary 3.1, since in G there exists a closed normal subgroup N of finite index such that $N \subset H$ holds]

<u>Remark:</u> The <u>induction principle</u> for groups in $\mathcal{L}$ is only true in quite restricted situations.

Is for example G a maximally almost periodic locally compact group with σ-compact commutator subgroup and H a compact normal subgroup of G, then with G/H also G is in $\mathcal{L}$.

[The proof of this result appearing already in [1], 321 relies on the method of Takahashi's in [17]]

As a consequence one concludes again that Takahashi groups are in $\mathcal{L}$. Without the assumption of maximal almost periodicity the induction principle is false as the following example of a locally compact not maximally almost periodic group G with compact commutator subgroup shows: Let G be the group

$$\left\{\begin{pmatrix} 1 & a & b \\ 0 & 1 & c \\ 0 & 0 & 1 \end{pmatrix} : a,b,c \in \mathbb{R}\right\} \Big/ \left\{\begin{pmatrix} 1 & 0 & n \\ 0 & 1 & 0 \\ 0 & 0 & 1 \end{pmatrix} : n \in \mathbb{Z}\right\}$$

Then the center

$$Z(G) = \left\{\begin{pmatrix} 1 & 0 & b \\ 0 & 1 & 0 \\ 0 & 0 & 1 \end{pmatrix} : b \in \mathbb{R}\right\} \Big/ \left\{\begin{pmatrix} 1 & 0 & n \\ 0 & 1 & 0 \\ 0 & 0 & 1 \end{pmatrix} : n \in \mathbb{Z}\right\}$$

of G is a compact normal subgroup of G (thus being in $\mathcal{L}$), G/Z(G) $\cong \mathbb{R}^2 \in \mathcal{L}$, but $G \notin \mathcal{L}$, since it is not maximally almost periodic.

4. Moore groups and groups with Chu duality

A locally compact group G will be called Moore group, if every irreducible (continuous, unitary) representation of G is necessariliy finite dimensional.

Let the classes of central, Takahashi and Moore groups be denoted by $\mathcal{Z}$, $\mathcal{T}$ and $\mathcal{M}$ resp.

Plainly all compact and all locally compact abelian groups are in $\mathcal{Z}$ and it is known that $\mathcal{Z} \subset \mathcal{T}$ (see [6]).

If we denote the class of all locally compact maximally almost periodic groups by $\mathfrak{M}$, we know by the Gelfand Raikov theorem that $\mathcal{M} \subset \mathfrak{M}$.

A characterization of the class $\mathcal{M}$ given by C. C. Moore in a yet unpublished paper [13] and a contribution of Robertson [15] yield the implications $\mathcal{T} \subset \mathcal{M}$ and $\mathcal{M} \subset \mathcal{L}$.

Theorem 4.1 (Moore) For any locally compact group G the following statements are equivalent:

(i) $G \in \mathcal{M}$

(ii) $G = \varprojlim_{\alpha \in A} G_\alpha$ with $G_\alpha \in \mathcal{M}$ and Lie for all $\alpha \in A$ and G_α is a finite extension of an open subgroup $H_\alpha \in \mathcal{Z}$ for every $\alpha \in A$.

[The structure of the proof is the following:

Let G be a locally compact group.

1. If H is an open subgroup of G of finite index, then $G \in \mathcal{M}$ iff $H \in \mathcal{M}$. Thus

2. If G has an open subgroup H of finite index which is in $\mathfrak{Z}$, then $G \in \mathfrak{M}$
3. If $G = \varprojlim_{\alpha \in A} G_\alpha$ with $G_\alpha \in \mathfrak{M}$ for all $\alpha \in A$, then $G \in \mathfrak{M}$
4. (hardest step to prove): If $G \in \mathfrak{M}$ and Lie, then there is an open subgroup of finite index in G which is in $\mathfrak{Z}$

Let $\mathfrak{S}$ be the class of locally compact groups which have arbitrarily small invariant neighoborhoods. Then

5. $\mathfrak{M} \subset \mathfrak{S}$
6. $G \in \mathfrak{S}$ iff $G = \varprojlim_{\alpha \in A} G_\alpha$ with $G_\alpha \in \mathfrak{S}$ and Lie for all $\alpha \in A$

(see theorem 2.11 in [7] and also a forthcoming paper by Wang [18]).

In particular the theorem implies that the class $\mathfrak{M}$ is closed under forming projective limits, (finite) direct products and finite extensions.

We proceed to a generalization of the theorem:

<u>Theorem 4.2</u> (Robertson) For any locally compact group G the following statements are equivalent:

(i) $G \in \mathfrak{M}$

(ii) G is a finite extension of a characteristic subgroup of G which is in $\mathfrak{T}$.

[1. $\mathfrak{T} \subset \mathfrak{S}$

This can be proved using 17.3.7 of [3]

An alternate proof follows at the end of the paragraph.

2. Let $G \in \mathfrak{M}$ and H be a closed normal subgroup of G such that G/H is abelian. Suppose in addition that H is a compactly generated abelian Lie group.

Then G is a finite extension of an abelian (normal sub) group.

From 2. one derives (using Bohr compactification):

3. Let $G \in \mathfrak{M}$ and H be a compact normal subgroup of G such that G/H is abelian. Suppose in addition that H is an abelian Lie group. Then G is a finite extension of an abelian group.

4. Any discrete $G \in \mathfrak{T}$ is a finite extension of an abelian group.

In fact by assumption K := K(G) is finite. For any $x \in G$ the stability subgroup C(x) of x w.r.t. the inner automorphisms of G has finite index in G. For the centralizer C(K) of K one has

$C(K) = \bigcap_{x \in K} C(x)$, thus C(K) has finite index in G.

Since $C(K) \in \mathfrak{T}$, one can replace G by C(K).

K(G) is abelian by $K(C(K)) \subset C(K) \cap K$.

Therefore 3. implies the result.

Let $\mathfrak{F}$ be the class of locally compact groups whose conjugacy classes are relatively compact.

5. To every $G \in \mathfrak{F}$ there is an exact sequence of locally compact groups

$$\{e\} \to K \to G \to \mathbb{R}^n \times D \to \{e\}$$

where K is a compact group, D is a discrete group in $\mathfrak{T}$ and $n \geq 1$

(For a proof see [12])

From 5. follows

6. Any $G \in \mathfrak{T}$ has the form $\mathbb{R}^n \times H$, where H_o is compact. (For a proof see [7], where a more general structure theorem is established)

7. Any Lie group $G \in \mathfrak{T}$ is in $\mathfrak{M}$.

In fact: By 6. we can assume that G_o is compact, and therefore G/G_o is a discrete group in $\mathfrak{T}$, thus by 4. a finite extension of an abelian group. Since $\mathcal{T}$ is closed under forming closed subgroups, we assume by 1. of 4.1 that G/G_o is abelian.

Let $N := K(G_o)$. Then by corollary 6.5 in [8], p. 122 N is a compact semi simple connected Lie group and we therefore suppose that G operates on N by inner automorphisms. Applying 3. to $H := G/N$ we can assume that H is abelian.

Let Z be the centralizer of N in G. Then Z is a closed normal subgroup of G and one has $G = NZ$. The isomorphism theorem yields

$$Z/_{Z \cap N} \simeq NZ/_N = H .$$

and so $Z/_{Z\cap N}$ is abelian.

$Z \cap N$ is a compact abelian Lie group and $Z \in \mathfrak{M}$. Thus we assume by 3. that Z is abelian. But then $Z \subset Z(G)$, the canonical mapping from N onto $G/_{Z(G)}$ is continuous and since N is compact, $G \in \mathcal{Z}$.

Therefore G is a finite extension of a group in $\mathcal{Z}$ and by theorem 4.1 $G \in \mathcal{M}$.

8. $\mathcal{T} \subset \mathcal{M}$

Let $G \in \mathcal{T}$. Then $G \in \mathcal{J}$ and by 6. of 4.1 there existes a family $(K_\alpha)_{\alpha \in A}$ of compact normal subgroups K_α of G such that $G/_{K_\alpha}$ is a Lie group for all $\alpha \in A$ and $\bigcap_{\alpha \in A} K_\alpha = \{e\}$.

Obviously $G/_{K_\alpha} \in \mathcal{T}$.

By 7. $G/_{K_\alpha} \in \mathcal{M}$ for all $\alpha \in A$ and by 3. of 4.1 $G \in \mathcal{M}$.

Finally we prove the <u>theorem</u>:

Let G be as in (ii). 8. and 1. of 4.1 yield the result.

Conversely let $G \in \mathfrak{U}$ (as in (i)). By theorem 4.1 there is a compact normal subgroup K of G, such that G/K is a finite extension of a group $H' \in \mathfrak{Z}$. But $\mathfrak{Z} \subset \mathfrak{T}$. Denoting the canonical mapping from G onto G/K by p and putting $H := p^{-1}(H')$ one obtains a subgroup H of G which is of finite index and in $\mathfrak{M}$.
By $K(H) \subset p^{-1}(K(H^{-1}))$ H is in $\mathfrak{T}$.
In addition H can be chosen characteristic by replacing H by $H_1 := \{x \in G : \mathrm{Int}(G)(x) \text{ is relatively compact}\}$. Then $H \subset H_1$ and H_1 is a characteristic subgroup of G which is in $\mathfrak{T}$.

In particular the theorem implies that $\mathfrak{U}$ is closed under forming closed subgroups.

Using theorems 3.3 and 4.2 we obtain immediately

Theorem 4.3 $\mathfrak{U} \subset \mathfrak{L}$

Collecting the structural properties of the classes introduced in this paragraph one finds the following inclusions:

First of all: $\mathfrak{Z} \subset \mathfrak{T} \subset \mathfrak{U} \subset \mathfrak{M}$

and also $\mathfrak{U} \subset \mathfrak{L} \subset \mathfrak{M}$,

but $\mathfrak{L} \neq \mathfrak{M}$

as the following example (of Roeders's) shows: Let p be a prime integer and A,B the groups $\mathbb{Z}_{p^2}$, $\mathbb{Z}_p$ with generators a,b resp. Every element in Aut(A) is given by $a \to a^n$ for some $n \geq 1$ with $(n,p) = 1$. Since Aut(A) has exactly $p(p-1)$ elements, there is a subgroup of Aut(A) of order p generated by $a \to a^r$ for $r \geq 1$.

We now define a non commutative metacyclic group G_p with generators a,b and relations

$$\begin{cases} b^{-1}ab = a^r \\ a^{p^2} = b^p = e \end{cases}$$

By [2] , 338 every irreducible representation of G_p which is not finite dimensional is an induced representation of a one dimensional representation of A and has therefore dimension $p = [G_p : A]$. It follows

$$G_p^{(p-1)} = G_p^{(1)} = K(G_p) \neq \{e\}$$

Putting $G := \prod^*_{p \text{ prime}} G_p$ and furnishing G with the discrete topology one obtains:

$G \in \mathfrak{M}$, but for any prime integer p

$$G^{(p-1)} \cap G_p^{(p-1)} = G_p^{(p-1)} \neq \{e\}$$

and thus by 3.3 (with $V := \{e\}$) $G \notin \mathfrak{L}$.

Furthermore $\mathfrak{L} \neq \mathfrak{U}$.

In fact there exists an example of a non Moore group which has Chu duality. The following example arose from a discussion with J. Liukkonen.

One considers the discrete groups $K := \prod^* \mathbb{Z}_3$ and $L := \prod^* \mathbb{Z}_2$ where the weak direct products are formed from countably many identical factors and defines the discrete semi direct product $G := K \times_\beta L$ with defining homomorphism $\beta : L \longrightarrow \mathrm{Aut}(K)$ given by

$$\beta((\varepsilon))((m)) := (\varepsilon m)$$

for all $(\varepsilon) \in L$ with $\varepsilon \in \{1,-1\}$ and $(m) \in K$ with $m \in \{0,1,2\}$.

A similar example is due to Murakami, see [9] , 190 .

G is in $\mathfrak{M}$, but not in $\mathfrak{U}$, which is easily deduced from Theorem 4.1 or 4.2, since $G \notin \mathfrak{Z}$.

In order to show that $G \in \mathcal{L}$ one proceeds as follows:
Let $G \in \mathfrak{M}$ and denote by $\tilde{G}$ the Bohr compactification of G and by $G^{\vee}$ the Chu quasi dual $\mathrm{Rep}(G)^{\vee}$ of G.
Let $G_1 := \overline{K(G)}$, $\tilde{G}_1 := \overline{K(\tilde{G})}$ and $G_1^{\vee} := \tilde{G}_1 \cap G^{\vee}$ $(\subset \tilde{G}_1 = (G_1)^{\sim})$.

One shows:

1. If $Q \in G_1^{\vee}$ and $D, D' \in \mathrm{Rep}(G)$ such that

$$\mathrm{Res}_{G_1} D = \mathrm{Res}_{G_1} D',$$

   then $Q(D) = Q(D')$
2. The mapping $\Omega_o := \varphi \circ \Omega$ composed of the Chu homomorphsm $\Omega : G \to G^{\vee}$ and of the canonical mapping $\varphi : G^{\vee} \to G^{\vee}/G_1^{\vee}$ is surjective.
3. Let now G be the above example $K \times_{\beta} L$. Then $G_1 := K$ and one shows $G_1^{\vee} \subset \Omega(G_1) = K$, thus obtains $\Omega(G) = G^{\vee}$ setwise.
4. Since G is discrete, one shows that also $G^{\vee}$ is discrete and therefore $G \in \mathcal{L}$.

Furthermore it should be noted that $\mathcal{L} \subset \mathcal{I}$.

[$\mathrm{Rep}(G)^{\vee}$ carries the topology $\mathcal{T}_c$, thus a basic neighborhood of the identity in $\mathrm{Rep}(G)^{\vee}$ is given by

$$N(K, \varepsilon) := \left\{ Q \in \mathrm{Rep}(G)^{\vee} : \| Q(D) - I_{n(D)} \| < \varepsilon \text{ for all } D \in K \right\},$$

where K is a compact subset of Rep(G) and $\varepsilon > 0$.
For any $Q \in \mathrm{Rep}(G)^{\vee}$, all $Q' \in N(K, \varepsilon)$ and $D \in K$ one has

$$\| Q Q' Q^{-1}(D) - I_{n(D)} \| \leq \| Q(D) \| \, \| Q'(D) - I_{n(D)} \| \, \| Q^{-1}(D) \|$$

$$= n(D) \| Q'(D) - I_{n(D)} \| < n(D) \varepsilon$$

and therefore $Q\, N(K,\varepsilon) Q^{-1} \subset N(K, n(D)\varepsilon)$ for all $Q \in Rep(G)^{\vee}$.]

But plainly one has $\mathcal{L} \neq \mathcal{S}$, since there are non maximally almost periodic groups which have small invariant neighborhoods.

Finally it follows $\mathcal{T} \subset \mathcal{S}$ a result which is basic for the proof of Theorem 4.2.

5. Roeder's characterization of Chu duality

Let G_1 and G_2 be two locally compact groups.

Definition: A continuous mapping ϕ from $Rep(G_1)$ into $Rep(G_2)$ is called a quasi homomorphism, if the following conditions are satisfied:

(QH1) $\phi(D) \in Rep_{n(D)}(G_2)$ for all $D \in Rep(G_1)$

(QH2) $\phi(D \oplus D') = \phi(D) \oplus \phi(D')$ for all $D, D' \in Rep(G_1)$

(QH3) $\phi(D \otimes D') = \phi(D) \otimes \phi(D')$ for all $D, D' \in Rep(G_1)$

(QH4) $\phi(U^{-1}DU) = U^{-1}\phi(D)U$ for all $D \in Rep(G_1)$ and $U \in \mathcal{U}(n(D))$.

The collection of all quasi homomorphisms from $Rep(G_1)$ into $Rep(G_2)$ will be denoted by $H(G_1,G_2)$ (= $Hom(Rep(G_1),Rep(G_2))$).

Furnished with the topology $\mathcal{T}_c$ $H(G_1,G_2)$ becomes a Hausdorff space.

For each $\phi \in H(G_1,G_2)$, $n \geq 1$ and subsets K of $Rep_n(G_1)$, F of G_2 and V of $\mathcal{U}(n)$ one defines

$$W(\phi;K,F,V) := \{\psi \in H(G_1,G_2) : \phi(D)(x^{-1})\psi(D)(x) \in V \text{ for all } D \in K \text{ and all } x \in F\}$$

Furthermore one puts

$$\mathcal{W}(\phi) := \{W(\phi; K, F, V) : K \text{ compact in } \mathrm{Rep}_n(G_1),$$
$$F \text{ compact in } G_2,\ V \text{ an open neighborhood}$$
$$\text{of the identity in } \mathfrak{U}(n) \text{ and } n \geq 1\}$$

and

$$\mathcal{W} := \{\mathcal{W}(\phi) : \phi \in H(G_1, G_2)\}.$$

Then it is proved in [14] that $\mathcal{W}$ is a base for the topology $\mathcal{T}_c$ in $H(G_1,G_2)$.

(See also [16] and [4] on this matter.)

Using this description of the topology in $H(G_1,G_2)$ one proceeds to a characterization of the class $\mathfrak{L}$: For (locally compact) groups G_1 and G_2 one defines a mapping ϕ_{G_1,G_2} from $\mathrm{Hom}(G_1,G_2)$ (furnished with the topology $\mathcal{T}_c$) into $H(G_2,G_1)$ by

$$\phi_{G_1,G_2}(f) := f^{\wedge}$$

for all $f \in \mathrm{Hom}(G_1,G_2)$, where $f^{\wedge}$ is defined by

$$f^{\wedge}(D) := D \circ f$$

for all $D \in \mathrm{Rep}(G_2)$.

For all $f \in \mathrm{Hom}(G_1,G_2)$ one has in fact $f^{\wedge} \in H(G_2,G_1)$.

<u>Theorem 3.1</u> (Roeder) Let G be a locally compact group.
Then $G \in \mathfrak{L}$ iff $\phi_{\mathbb{Z},G}$ is a homeomorphism, i.e. iff $\mathrm{Hom}(\mathbb{Z},G)$ and $H(G,\mathbb{Z})$ are homeomorphic spaces.

[(i) Let $G \in \mathcal{L}$. One verifies immediately that the mapping $f \to f^\wedge$ from $\mathrm{Hom}(\mathbb{Z},G)$ into $H(G,\mathbb{Z})$ is injective (since $\mathcal{L} \subset \mathcal{M}$) and surjective (since $G \in \mathcal{L}$). Since G is a topological group, the space $\mathcal{C}(\mathbb{Z},G)$ of all continuous mappings from $\mathbb{Z}$ into G becomes a topological group, if one introduces pointwise multiplication and the topology $\mathcal{T}_c$.

Since $\mathrm{Hom}(\mathbb{Z},G)$ is a subspace of $\mathcal{C}(\mathbb{Z},G)$, an open neighborhood of an element $f \in \mathrm{Hom}(\mathbb{Z},G)$ has the form

$$X(f;F,V) := \left\{ g \in \mathrm{Hom}(\mathbb{Z},G) : f(x)^{-1}g(x) \in V \text{ for all } x \in F \right\}$$

for compact subsets F of $\mathbb{Z}$ and open neighborhoods V of the identity e in G.

But now G has Chu duality (the Chu duality being denoted by Ω), thus the topology of G can be described in terms of the topology $\mathcal{T}_c$ on $\mathrm{Rep}(G)^\vee$.

A neighborhood system of the identity of G is given by the family

$$\left\{ \Omega^{-1}(N(K,V)) : K \text{ compact in } \mathrm{Rep}_n(G),\ V \text{ an open neighborhood of the identity of } \mathcal{U}(n) \text{ and } n \geq 1 \right\}.$$

Therefore a neighborhood system of any $f \in \mathrm{Hom}(\mathbb{Z},G)$ is given by the family

$$\left\{ X(f;F,\Omega^{-1}(N(K,V))) : F,K,V \text{ as above} \right\}.$$

Plainly

$$g \in X(f;F,\Omega^{-1}(N(K,V))) \text{ iff } g^\wedge \in W(f^\wedge;K,F,V)$$

for all F,K,V as above, i.e. the mapping $\phi_{\mathbb{Z},G}$ is a bijection transforming neighborhoods of elements of $\mathrm{Hom}(\mathbb{Z},G)$ onto neighborhoods of elements of $H(G,\mathbb{Z})$ and conversely, thus $\phi_{\mathbb{Z},G}$ is a homeomorphism from $\mathrm{Hom}(\mathbb{Z},G)$ onto $H(G,\mathbb{Z})$.

(ii) Evidently $\mathrm{Rep}(\mathbb{Z}) = \bigcup_{n \geq 1} \mathrm{Hom}(\mathbb{Z}, \mathfrak{U}(n))$ (considered as the topological union of the spaces $\mathrm{Hom}(\mathbb{Z}, \mathfrak{U}(n))$ for $n \geq 1$) and therefore

$$\mathrm{Rep}(\mathbb{Z}) \cong \mathfrak{U} = \bigcup_{n \geq 1} \mathfrak{U}(n).$$

Thus one has a homeomorphism ϱ from $\mathrm{Rep}(\mathbb{Z})$ onto $\mathfrak{U}$ defined by $\varrho(D) := D(1)$ for all $D \in \mathrm{Rep}(\mathbb{Z})$ with the properties

(a) $\varrho(D \oplus D') = \varrho(D) \oplus \varrho(D')$ for all $D, D' \in \mathrm{Rep}(\mathbb{Z})$

(b) $\varrho(D \otimes D') = \varrho(D) \otimes \varrho(D')$ for all $D, D' \in \mathrm{Rep}(\mathbb{Z})$

(c) $\varrho(U^{-1}DU) = U^{-1}\varrho(D)U$ for all $D \in \mathrm{Rep}(\mathbb{Z})$

and all $U \in \mathfrak{U}(n(D))$.

We define the mapping $\sigma := \sigma_\varrho$ from $H(G, \mathbb{Z})$ into $\mathrm{Rep}(G)^\vee$ by

$$\sigma(\phi) := \varrho \circ \phi$$

for all $\phi \in H(G, \mathbb{Z})$.

By assumption $\phi_{\mathbb{Z},G}$ is a homeomorphism from $\mathrm{Hom}(\mathbb{Z}, G)$ onto $H(G, \mathbb{Z})$.

We further define the mapping τ from G into $\mathrm{Hom}(\mathbb{Z}, G)$ by

$$\tau(x)(1) := x$$

for all $x \in G$.

Of course τ is a homeomorphism from G onto $\mathrm{Hom}(\mathbb{Z}, G)$.

Since $\Omega = \sigma \circ \phi_{\mathbb{Z},G} \circ \tau$, Ω is a topological isomorphism.

<u>Remark:</u> More generally it can be shown that for a group $G_2 \in \mathcal{L}$ and any locally compact group G_1 the spaces $\mathrm{Hom}(G_1, G_2)$ and $H(G_2, G_1)$ are homeomorphic.

A detailed analysis of those locally compact groups G, for which $\mathrm{Hom}(\mathbb{Z},G) \cong H(G,\mathbb{Z})$ holds, will possibly yield further properties of the class $\mathfrak{L}$.

References

[1] H. Chu, Compactification and duality of topological groups
Trans AMS 123(1966), 310-324

[2] C. W. Curtis, I. Reiner, Representation Theory of Finite Groups and Associative Algebras
Interscience (1962)

[3] J. Dixmier, Les C* -algèbres et leurs représentations
Gauthier-Villars (1964)

[4] J. Ernest, Notes on the duality theorem of non-commutative non compact topological groups
Tôhoku Math. J. 16(1964), 291-296

[5] M. Goto, Note on a topology of a dual space
Proc. AMS 12(1961), 41-46

[6] S. Grosser, M. Moskowitz, On central topological groups
Trans AMS 127(1967), 317-340

[7] S. Grosser, M. Moskowitz, Compactness conditions in topological groups I, II
(to appear in J. Reine Angew. Math.)

[8] S. Helgason, Differential Geometry and Symmetric Spaces
Academic Press (1962)

[9] H. Heyer, Dualität lokalkompakter Gruppen
Lecture Notes in Mathematics 150, Springer (1970)

[10] E. Hewitt, K. A. Ross, Abstract Harmonic Analysis I, II
Springer (1963/1970)

[11] E. Joachim, Dualitätstheorie der nicht-kommutativen
Gruppen mit Normalteilertopologie
Math. Z. 111(1969), 46-52

[12] J. R. Liukonnen, Compactness conditions and duals of
topological groups
PhD thesis Columbia University (1970)

[13] C. C. Moore, Groups with finite dimensional irreducible
representations
(yet unpublished)

[14] D. W. Roeder, A characterization of unitary duality
Trans AMS 148(1970), 129-135

[15] L. C. Robertson, A note on the structure of Moore groups
Bull AMS 75(1969), 594-599

[16] K. Suzuki, Notes on the duality theorem of non-commutative
topological groups
Tôhoku Math. J. 15(1963), 182-186

[17] S. Takahashi, A duality theorem for representable locally
compact groups with compact commutator subgroup
Tôhoku Math. J. 5(1953), 115-121

[18] S. P. Wang, A note on small invariant neighborhoods of the
identity
(to appear in Proc. AMS)

INVARIANCE OF DECISION FUNCTIONS UNDER LIE GROUPS I

M. Behara

McMaster University
and
University of Karlsruhe

As in [1], let the measurable spaces $(\Theta, \mathcal{A})$, $(\Omega, \mathcal{B})$ and $(\Delta, \mathcal{D})$ denote the spaces of states of nature, samples, and decisions. Let us assume Θ, Ω and Δ to be subsets of n-dimensional enclidean spaces in which coordinate systems are introduced. Let $\mathcal{J}^1$ (resp. $\mathcal{J}^2$ and $\mathcal{J}^3$) be a group of measurable transformations T^1 (resp. T^2 and T^3) acting on Θ (resp. Ω and Δ) such that the image of $(T^1, \theta) \in \mathcal{J}^1 \times \Theta$ (resp. $(T^2, \omega) \in \mathcal{J}^2 \times \Omega$ and $(T^3, \delta) \in \mathcal{J}^3 \times \Delta$) is unique point $T^1\theta = \theta'$ (resp. $T^2\omega = \omega' \in \Omega$ and $T^3\delta = \delta' \in \Delta$).

Supposing again that each transformation T^1 (resp. T^2 and T^3) is determined by a finite number of parameters, we have $\theta' = T^1_{a_1}\theta$ (resp. $\omega' = T^2_{a_2}\omega$ and $\delta' = T^3_{a_3}\delta$) where a_1 (resp. a_2 and a_3) is a point in r-dimensional parameter space α_1 (resp. α_2 and α_3) which is endowed with coordinate system similar to above.

This article is a part of lectures delivered at the University of Karlsruhe during the summer of 1968 when the author held a visiting professorship there. The author expresses his deep gratitude to Prof. D. Bierlein for his kind invitation and for many fruitful discussions.

In detailed notations we have

$$\theta_i' = T^{1_i}_{(a_{1_1}, \ldots a_{1_r})} (\theta_1, \ldots, \theta_n)$$

$$\omega_i' = T^{2_i}_{(a_{2_1}, \ldots a_{2_r})} (\omega_1, \ldots, \omega_n)$$

$$\delta_i' = T^{3_i}_{(a_{3_1}, \ldots a_{3_r})} (\delta_1, \ldots, \delta_n)$$

for $i = 1, \ldots , n$.

Let us assume that the transformations T^{1_i} are analytic with respect to θ_{1_i} and a_{1_j}, $i = 1, \ldots , n$; $j = 1, \ldots , r$. Similarly, we assume the analyticity of T^{2_i} and T^{3_i}. The group $\mathfrak{J}^1$ of transformations T^1, or, more specifically, T^{1_i} is then a Lie group with underlying manifold Θ. Similarly, $\mathfrak{J}^2$ and $\mathfrak{J}^3$ are Lie groups over the manifolds Ω and Δ respectively. Details of Lie groups may be seen in [4], [5] and [8].

The product $\mathfrak{J}$ of the Lie groups $\mathfrak{J}^1$, $\mathfrak{J}^2$ and $\mathfrak{J}^3$ is also a Lie group (see [4], p. 130). Let us denote the underlying product manifold $\Theta \times \Omega \times \Delta$ of $\mathfrak{J}$ by M. $\mathfrak{J} \times M \to M$ gives the image of $(T,m) \in \mathfrak{J} \times M$ in M uniquely at $Tm = m'$. We note that, even if the manifolds $(\Theta \times \Omega) \times \Delta$, $\Theta \times (\Omega \times \Delta)$ and $\Theta \times \Omega \times \Delta$ are not the same, there is a natural analytic isomorphism between any two of them. ([4], p. 76).

Let us also note that the transformations corresponding to points of the parameter spaces α_1, α_2 and α_3 form what are known as parameter groups $\mathfrak{J}_{a_1}$, $\mathfrak{J}_{a_2}$ and $\mathfrak{J}_{a_3}$ of the groups $\mathfrak{J}^1$, $\mathfrak{J}^2$ and $\mathfrak{J}^3$ respectively. Denote the parameter group of $\mathfrak{J}$ by $\mathfrak{J}_a$.

The statistical decision problem described in [1] remains

invariant under the group $\mathfrak{J}$ if

(i) the transition probability $P = \{P(\theta,B)\}$ relative to $(\Theta, \mathcal{A})$ and $(\Omega, \mathcal{B})$ remains invariant under $\mathfrak{J}$.

(ii) The randomized decision function $S = \{S(\omega, D)\}$ relative to $(\Omega, \mathcal{B})$ and $(\Delta, \mathcal{D})$ remains invariant under $\mathfrak{J}$.

(iii) The randomized loss $L = \{L(\theta, D)\}$ relative to $(\Theta, \mathcal{A})$ and $(\Delta, \mathcal{D})$ remains invariant under $\mathfrak{J}$.

Let us now consider the group $\mathfrak{J}'$ and its infinitesimal transformations. Let the transformation with $a_{1_1} = 0, \ldots a_{1_r} = 0$ be the identity of $\mathfrak{J}'$. Giving now infinitesimal values of the form $\varepsilon_j(a_1, da_1)$ to the parameters a_{1_j} we get the infinitesimal transformation of the group $\mathfrak{J}'$ by

(1) $$\theta_i' = T_\varepsilon \theta_i = \theta_i + \sum_{j=1}^{r} \left(\frac{\partial T^{1_i}}{\partial a_{1_j}}\right)_0 \varepsilon_j(a_1, da_1)$$

$$= \theta_i + \delta\theta_i \text{ (say)}, \; i = 1, \ldots, n$$

(neglecting the powers of $\delta\theta_i$ higher than the first). Note that the infinitesimal generator of the group $\mathfrak{J}$ given by

(2) $$Y_j = \sum_i \left(\frac{\partial T^i}{\partial a_j}\right)_{a=e} \frac{\partial}{\partial m_i}$$

is the sum of the infinitesimal generators of $\mathfrak{J}^1$, $\mathfrak{J}^2$ and $\mathfrak{J}^3$ where a_j, $j = 1, \ldots, r$ is the jth coordinate of the parameter space α of the parameter group $\mathfrak{J}_a$ of $\mathfrak{J}$ whose identity is given at $a = e$. For a general differentiable function F defined on the manifold M, and setting $\left(\frac{\partial T^i}{\partial a_j}\right)_{a=e} = \xi_{ij}$ we get an infinitesimal transformation denoted by

(2) $$Y_jF = \sum_i \xi_{ij} \frac{\partial F}{\partial m_i}.$$

The infinitesimal transformation T_ε applied to m_i may now be written as

(3) $$\delta F = \sum_j Y_jF \quad \varepsilon_j(a,da).$$

Let us now assume the loss function ℓ as well as the functions p and s to be general differentiable functions as described above. Then we have the following

Theorem 1. If the decision problem is invariant under the Lie group $\mathfrak{J}$, then the infinitesimal transformation

(4) $$\delta\ell = \sum_j Y_j\ell \; \varepsilon_j(a,da) = 0.$$

(5) $$\delta p = \sum_j Y_jp \; \varepsilon_j(a,da) = 0.$$

(6) $$\delta s = \sum_j Y_js \; \varepsilon_j(a,da) = 0.$$

Proof. Since the functions ℓ, p and s are independent of the set of coordinates a_j, $j = 1, \ldots, r$ of the parameter space α of $\mathfrak{J}_a$ of $\mathfrak{J}$ we have $\frac{\partial \ell}{\partial a_j} = \frac{\partial p}{\partial a_j} = \frac{\partial s}{\partial a_j} = 0$. Now from (1) and (3) we obtain the desired result. ▯

The coordinate system introduced earlier may be replaced by the equivalent concept of "frame" due to E. Cartan. A transformation T_a transforms an original frame R_o into a new frame R_a. The notion of infinitesimal transformation that moves a frame R_a to R_{a+da} is given by $T_a^{-1}\,T_{a+da}$. An equivalent form of (1) is given by

$$(7) \qquad \theta_i' = T_a^{-1}\,T_{a+da}\,\theta_i$$

$$= T_\gamma\,\theta_i \qquad = \theta_i + \sum_{j=1}^{r} \left(\frac{\partial T^{1i}}{\partial a_{1_j}}\right) \gamma_j(a_1, da_1).$$

The infinitesimal values $\gamma_j(a,da)$ (known as the Pfaffian forms) are called the relative components of R_a (or of the group).

We will now state some properties (see [8]) of the relative components.

1. $\gamma_j(a,da)$ are independent forms,
2. $\gamma_j(a,da)$ of a moving frame are invariant with respect to the parameter group,
3. The Pfaffian form invariant under the parameter group is uniquely represented by

$$(8) \qquad \Gamma(a,da) = \sum_{j=1}^{r} A_j(a)\,\gamma_j(a,da)$$

where A_j's are constants.

4. If the system

$$(9) \qquad \Gamma_i = 0 \qquad i = 1, \ldots, h$$

formed by h independent Pfaffian forms $\Gamma_i(a,da)$ is completely integrable and is invariant under the parameter group then it is equivalent to a system

$$(10) \qquad \Gamma_i^* = 0 \qquad i = 1, \ldots, h,$$

where each Γ_i^* is itself invariant under the parameter group.

Let the subgroup of transformations of $\mathfrak{J}_{a3}$ under which a fixed number of elements δ's of Δ remain invariant be denoted by $\mathfrak{J}'_{a_3}$ and let $\mathfrak{J}'_{a_3}$ be represented by a variety of dimension r-h ($h<r$) in the parameter space α_3 of $\mathfrak{J}_{a_3}$. (We use obviously similar notations for the spaces Ω and Θ and for the corresponding parameter groups etc.) Let the h-dimensional parameter space $\alpha'_3(z)$ be the homogenous space $\mathfrak{J}_{a_3}/\mathfrak{J}'_{a_3}$. A point z of the parameter space $\alpha'_3(z)$ has $z_1, \ldots, z_h$ coordinates. To each point z of $\alpha'_3(z)$ there corresponds an element δ of Δ. Thus, finding the invariant decision measure S_ω for the elements δ's of Δ and the invariant measure for the points z of $\alpha'_3(z)$ amounts to the same thing.

Let D be a set of points δ and let the invariant (under the r-parameter Lie group $\mathfrak{J}_{a_3}$) decision measure S_ω be of the form

(11) $\int_D f(\delta)\, d\delta$

But, since the system (9) is equivalent to

(12) $d\delta_1 = 0, \ldots, d\delta_h = 0,$

We must have, by (8)

(13) $d\delta_k = \sum_{j=1}^{h} A_{kj}(a)\, \gamma_j(a,da), \; k = 1, \ldots, h$

and hence

(14) $[d\delta_1, \ldots, d\delta_h] = \mathrm{Det}\,(A_{kj}(a))\, [\gamma_1, \ldots, \gamma_h]$

where square bracket [ ] denotes the exterior product.

Substituting (14) in (11) and requiring S_ω to be invariant for all set D of δ's we obtain, by the homogeneity of $\mathfrak{J}_{a_3}/\mathfrak{J}'_{a_3}$

(15) $f(\delta)\, \mathrm{Det}\,(A_{kj}(a))$ = a constant with respect to a.

The above result, due to S.S. Chern (See [8]) is stated below in the present frame work.

Theorem 2. The decision measure S and the loss measure L_θ (the probability measure P_θ) are invariant under the r-parameter

Lie group $\mathcal{J}_{a_3}(\mathcal{J}_{a_2})$ iff (15) holds. If the subgroup $\mathcal{J}'_{a_3}(\mathcal{J}'_{a_2})$ of $\mathcal{J}_{a_3}(\mathcal{J}_{a_2})$ reduces to an identity, the decision and loss measures are always invariant.

In order to be able to apply the results described herein as well as their consequences to certain decision problems given in [2], [6], [7], [9], [10], a brief introduction is in order. Let the spaces Θ, Ω and Δ break up into orbits under the influences of $\mathcal{J}^1$, $\mathcal{J}^2$ and $\mathcal{J}^3$ respectively. Now the orbits, rather than the points of the spaces Θ, Ω and Δ are considered. Invariance of the risk function would, for example, mean that it remains constant on orbits of Θ for an invariant decision function s. Finally, solution of these decision problems under certain decision criteria, for example, minimax criterion or Bayesified minimax criterion (see [10]) may be worked out.

REFERENCES

[1] M. Behara (1973): Statistical decision functions on metric spaces. To appear in Selecta Statistica Canadiana, vol. 3.

[2] M. Behara and N. Giri (1971): Locally and asymptotically minimax tests of some multivariate decision problems. Archiv der Mathematik. XXII.

[3] D.R. Brillinger (1963): Necessary and sufficient conditions for a statistical problem to be invariant under a Lie Group. Ann. Math. Statist. 34,p. 492-500.

[4] C. Chevalley (1946): Theory of Lie Groups. Princeton University Press.

[5] E.J. Hannan (1965): Group Representations and Applied Probability. Methuen and Co. Ltd. London.

[6] J. Kiefer (1957): Invariance, minimax sequential estimation, and continous time process. Ann. Math. Statist. 28, p. 573-601.

[7] E.L. Lehmann (1959): Testing Statistical Hypotheses. John Wiley, New York and London.

[8] L.A. Santalo (1953): Introduction to Inte ral Geometry. Hermann and Cie, Paris.

[9] M. Stone and R. von Randow (1968): Statistically inspired conditions on the group structure of invariant experiments. Zeitschrift für Wahrscheinlichkeitstheorie u. verw. Geb. 10, p. 70-80.

[10] O. Wesler (1959): Invariance theory and a modified minimax principle. Ann. Math. Statist. 30, p. 1-20.